D0174737

Loonshots

Loonshots

HOW to NURTURE the CRAZY IDEAS THAT WIN WARS, CURE DISEASES, and TRANSFORM INDUSTRIES

SAFI BAHCALL

ST. MARTIN'S PRESS NEW YORK

 Printed in the United States of America. For information, address St. Martin's Press, 175 Fifth Avenue, New York, NY 10010.

www.stmartins.com

Library of Congress Cataloging-in-Publication Data is available upon request.

ISBN 978-1-250-18596-9 (hardcover)
ISBN 978-1-250-22561-0 (International, sold outside the U.S., subject to rights availability)
ISBN 978-1-250-18597-6 (ebook)

Our books may be purchased in bulk for promotional, educational, or business use. Please contact your local bookseller or the Macmillan Corporate and Premium Sales Department at 1-800-221-7945, extension 5442, or by email at MacmillanSpecialMarkets@macmillan.com.

10 9

For my father,
John Bahcall,
who showed me and so many others
how to hold truth near and persevere

CONTENTS

Moonshot: (1) The launching of a spacecraft to the moon; (2) an ambitious and expensive goal, widely expected to have great significance.

Loonshot: A neglected project, widely dismissed, its champion written off as unhinged.

Loonshots

Prologue

A dozen or so years ago, a friend took me to see a play called *The Complete Works of William Shakespeare (Abridged)*. Three actors covered 37 plays in 97 minutes (including *Hamlet* in 43 seconds). They skipped the boring stuff. Not long afterward I was invited to give a talk at a business gathering. The topic was my choice, but it could not be related to my job. I presented "3,000 years of physics in 45 minutes"—the eight greatest ideas in the history of the field. I skipped the boring stuff.

That greatest hits show ran on and off until 2011, when the personal hobby crossed paths with a professional assignment. I was asked to join a group developing recommendations for the president on the future of US national research. On the first day, our chairman announced our mission. What should the president do to ensure that national research continues to improve the well-being and security of our country for the next fifty years? Our task, he said, was to create the next generation of the Vannevar Bush report.

Unfortunately, I'd never heard of Vannevar Bush, or his report. I soon learned that Bush developed a new system, during the Second World War, for nurturing radical breakthroughs astonishingly fast. His system helped the Allies win that war, and the United States lead the world in science and technology ever since. Bush's goal: that the US should be the initiator, not the victim, of innovative surprise.

What Bush did, and why he did it, came right back to one of those eight greatest ideas of physics: phase transitions.

In this book, I'll show you how the science of phase transitions suggests a surprising new way of thinking about the world around us—about the mysteries of group behavior. We will see why good teams will kill great ideas, why the wisdom of crowds becomes the tyranny of crowds when the stakes are high, and why the answers to these questions can be found in a glass of water.

I'll describe the science briefly (skipping the boring stuff). And then we'll see how small changes in *structure*, rather than *culture*, can transform the behavior of groups, the same way a small change in temperature can transform rigid ice to flowing water. Which will give all of us the tools to become the initiators, not the victims, of innovative surprise.

Along the way, you will learn how chickens saved millions of lives, what James Bond and Lipitor have in common, and where Isaac Newton and Steve Jobs got their ideas.

I've always appreciated authors who explain their points simply, right up front. So here's the argument in brief:

1. The most important breakthroughs come from *loonshots*, widely dismissed ideas whose champions are often written off as crazy.
2. Large groups of people are needed to translate those breakthroughs into technologies that win wars, products that save lives, or strategies that change industries.
3. Applying the science of *phase transitions* to the behavior of teams, companies, or any group with a mission provides practical rules for nurturing loonshots faster and better.

In thinking about the behavior of large groups of people in this way, we are joining a growing movement in science. Over the past decade, researchers have been applying the tools and techniques of phase transitions to understand how birds flock, fish swim, brains work, people vote, criminals behave, ideas spread, diseases erupt, and ecosystems collapse. If twentieth-century science was shaped by the search for fundamental laws, like quantum mechanics and gravity, the twenty-first will be shaped by this new kind of science.

None of which changes the well-established fact that physics rarely mixes with the study of human behavior, let alone sits down for a full-course meal, so some sort of explanation is in order. I was born into the field. Both my parents were scientists, and I followed them into the family business. After a few years, like many who follow their elders, I decided I should see other parts of the world. To my parents' horror, I chose the business world. They responded to my lost academic career with the five stages of grief, starting with denial (telling family friends it was just a phase), skipping quickly past anger to bargaining and depression, before settling into resigned acceptance. I missed science enough, however, that eventually I joined forces with a handful of biologists and chemists to start a biotech company developing new cancer drugs.

My interest in the strange behaviors of large groups of people began shortly afterward, with a visit to a hospital.

Introduction

One winter morning in 2003, I drove to the Beth Israel Deaconess Medical Center in Boston to meet a patient named Alex. Alex was 33, with the strong, graceful build of an athlete. He had been diagnosed with an aggressive form of cancer called Kaposi's sarcoma. Six regimens of chemotherapy had not stopped his disease. His prognosis was poor. A handful of scientists and I had spent two years preparing for this moment. Alex was scheduled to be the first patient to receive our new drug for treating cancer.

When I entered his room, Alex was lying in bed, attached to an IV drip, speaking softly to a nurse. A yellowish liquid, our drug, fed slowly into his arm. The physician had just left. Then the nurse, who had been writing up notes in the corner, closed her folder, waved, and left. Alex turned to me with a gentle smile and quizzical look. The frenzy of activity to get to this day—licensing discussions, financings, laboratory studies, safety experiments, manufacturing checks, FDA filings, protocol drafting, and years of research—melted away. Alex's eyes asked the only thing that mattered: would the yellowish liquid save his life?

Physicians see this look all the time. I didn't.

I pulled up a chair. We talked for nearly two hours, as the drug dripped into Alex's arm. Restaurants, sports, the best cycling paths in Boston. Toward the end, after a pause, Alex asked me what would be next, if our drug didn't work. I stumbled through some non-answer. But we both knew. Despite tens of billions of dollars spent every year on research by national labs and large research companies, sarcoma treatment hadn't changed in decades. Our drug was a last resort.

Two years later, I found myself pulling up a chair next to another bed, in a different hospital. My father had developed an aggressive type of leukemia. One older physician told me, sadly, that all he could offer was the same chemotherapy he had prescribed as a resident forty years earlier. Second, third, and fourth opinions and dozens of desperate phone calls confirmed what he said. No new drugs. Not even any promising clinical trials.

There are some technical reasons why cancer drug development is so difficult. So many things have broken down inside a cancer cell by the time it starts proliferating that there's no easy fix. Laboratory models are notoriously bad at predicting results in patients, which leads to high failure rates. Clinical trials take years to conduct and can cost hundreds of millions of dollars. All these points are true. But there's more.

MILLER'S PIRANHA

"They looked at me like I was a lunatic," Richard Miller told me.

Miller, an affable oncologist in his sixties, was explaining to me the reactions of research teams at large pharma companies to his suggestion of treating cancer patients with a new drug he had been working on. It was a chemical designed originally just for laboratory use, for experiments—a tool, like bleach.

Most drugs work by gently attaching themselves to the overactive proteins inside cells that trigger disease. Those proteins act like an army of hypercharged robots, causing cells to go haywire. The cells may start multiplying out of control, like cancer. Or they may attack the body's own tissues, like in severe arthritis. By attaching to the overactive proteins, drugs dial down their activity, quieting the cells, restoring order in the body.

Miller's drug, however, didn't gently attach; it was a piranha (irreversible binder, to chemists). It grabbed hold and never let go. The problem with piranhas is that you can't wash them out of your system if something goes wrong. If they latch on to the wrong protein, for example, they can cause serious, even fatal, toxicities. You don't give piranhas to patients.

Miller was the CEO of a struggling biotech company. Its first project, developed a decade before Miller's new drug, hadn't panned out. The company's stock price had fallen below a dollar, and it received a delisting notice from Nasdaq, meaning that it would soon be banished from the market for serious companies and transferred to the purgatory of flaky has-beens.

I asked Miller why he persisted with the piranha in that precarious state and despite so many rejections, even ridicule. Miller said he understood all the arguments against his drug. But there was a flip side: the drug was so strong that he could give a very low dose. Miller also served part-time as a physician at Stanford University. He explained that he knew his patients. Many had only months to live, were desperately looking for options, and understood the risks. The potential, in this context, justified the risk.

"There's a quote from Francis Crick that I love," Miller said. Crick was awarded the Nobel Prize for discovering, along with James Watson, the double-helix structure of DNA. "When asked what it takes to win a Nobel Prize, Crick said, 'Oh it's very simple. My secret had been I know what to ignore.'"

Miller shared the early laboratory results from his piranha with a handful of physicians, who agreed to proceed with a clinical trial in patients with advanced leukemias. But Miller's investors were not convinced. (Miller: "To this day, if you ask them [how the drug works], they wouldn't know.") He lost a boardroom battle and resigned as CEO.

The trial, however, continued. Not long after Miller left, early results came back. They were encouraging. The company began a much larger, pivotal study. Half the patients would receive standard therapy, half the new drug. In January 2014, the physicians monitoring that study, which enrolled nearly four hundred patients, recommended that the trial be stopped. The results were so spectacular—a nearly *ten times* higher response rate in patients who received Miller's drug, called ibrutinib, than in patients who received standard therapy—that denying patients in the control group access to ibrutinib was considered unethical.

The FDA approved the drug shortly afterward. A few months later, Miller's company, called Pharmacyclics, was acquired by one of those large pharma companies that had ridiculed the idea.

The price: $21 billion.

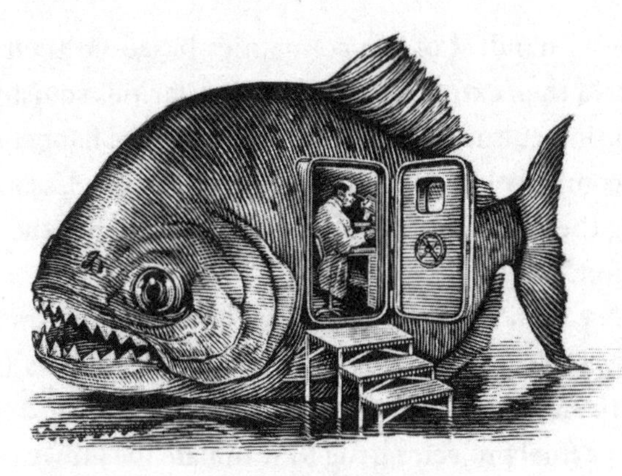

Scientist and piranha

Miller's piranha was a classic *loonshot.* The most important breakthroughs rarely follow blaring trumpets and a red carpet, with central authorities offering overflowing pots of tools and money. They are surprisingly fragile. They pass through long dark tunnels of skepticism and uncertainty, crushed or neglected, their champions often dismissed as crazy—or just plain dismissed, like Miller.

* * *

Drugs that save lives, like technologies that transform industries, often begin with lone inventors championing crazy ideas. But large groups of people are needed to translate those ideas into products that work. When teams with the means to develop those ideas reject them, as every large research organization rejected Miller's piranha, those breakthroughs remain buried inside labs or trapped underneath the rubble of failed companies.

Miller just barely saved his idea. Most loonshots never get the chance.

There's something at the core of how large groups behave that we just don't understand, despite the mountains of mind-numbing print written on the subject. Every year, glossy magazines celebrate the winning cultures of innovative teams. Covers show smiling employees raising gleaming new products like runners raising the Olympic torch. Leaders reveal their secrets. And then, so often, those companies crash and burn. The people are the same; the culture is the same; yet seemingly overnight, they turn. Why?

Articles and books on *culture* have always felt squishy to me. I hear culture, I think yogurt. For example, one popular book, typical of the genre, identifies a handful of top companies based on their stock price performance and then extracts from their similarities squishy lessons on creating a winning culture. One of those companies happens to be Amgen, a biotech company I know well. Among the Amgen lessons extracted: "By embracing the myriad of possible dangers, they put themselves in a superior position."

The real story with Amgen is that after a couple of years in business, the company was nearly bankrupt, all its initial projects (including a chicken growth hormone and pig vaccines) had failed, and time was running out on a final project, a drug to stimulate the growth of red blood cells. A handful of companies were pursuing the same goal. Amgen got to the finish line just ahead of its competitors. Much of that was due to a University of Chicago professor named Eugene Goldwasser. Goldwasser had worked on the problem for twenty years and held the key to winning the race: an eight-milligram vial of purified protein, painstakingly extracted from 2,550 liters of human urine. The purified protein contained the code to making the drug. He decided to give that vial to Amgen rather than its main competitor, Biogen. Biogen's CEO had refused to pick up the check for dinner one night.

The drug, called erythropoietin, or epo for short, turned out to be far, far more successful than anyone, including Amgen, imagined—eventually bringing in $10 billion a year. Amgen had won the drug-discovery lottery. Once it had the drug, Amgen sued everyone else in the business (including its partner, Johnson & Johnson, which had saved Amgen when it was struggling) to stop them from competing. For the next fifteen years, Amgen was unable to repeat its drug-discovery success. Its poor research output, as measured by number of patents awarded, was noted by the culture-analyzing book, which concluded that being "innovative doesn't seem to matter very much."

Amgen may not have had good research, but it did have good lawyers. It won every lawsuit, and its competitors gave up. Among insiders, the company was called "a law firm with a drug."

Useful lessons from Amgen's story include picking up the check for dinner and hiring good lawyers. But otherwise, extracting culture tips, after the fact, from its terrific stock price performance is like asking the guy who just won the lotto to describe the socks he was wearing when he bought the winning ticket.

My resistance to after-the-fact analyses of culture comes from being trained as a physicist. In physics, you identify clues that reveal fundamental truths. You build models and see if they can explain the world around you. And that's what we will do in this book. We will see why *structure* may matter more than *culture*.

* * *

After a few months of treatment at Beth Israel, Alex recovered. He is still alive today, as I write this.* My father did not recover. No treatment I could find, none of the desperate phone calls, none of the expert friends and colleagues, none of the work I'd done, nothing made any difference. He died a few months after he was diagnosed, but for many years after, I felt I was still fighting that battle, that if I worked hard enough, I could find something for what he had, and it would matter. That I would stop feeling I had let him down. In a recurring dream, I hand the nurse by his bed a vial. She plugs it into his IV. Disease disappears.

Dozens of promising drug candidates for treating my father's condition were buried then. They remain buried today.

To liberate those buried drugs and other valuable products and technologies, we need to begin by understanding why good teams, with the best intentions and excellent people, kill great ideas.

WHEN TEAMS TURN

In the 1970s, Nokia was an industrial conglomerate famous mostly for its rubber boots and toilet paper. Over the next two decades, it would pioneer the first cellular network, the first car phone, the first all-network analog phone, and the first wildly successful GSM phone. By the early 2000s, it was selling *half* the smartphones on the planet. It became, briefly, the most valuable company in Europe. A *BusinessWeek* cover story declared, "Nokia has become synonymous with success." *Fortune* revealed Nokia's secret: it was "the least hierarchical big company in the world." The CEO explained that the key was the culture: "You are allowed to have a bit of fun, to think unlike the norm . . . to make a mistake."

---

* Alex's name has been changed. For more on his treatment, see the notes.

In 2004, a handful of excited Nokia engineers created a new kind of phone: internet-ready, with a big color touchscreen display and a high-resolution camera. They proposed another crazy idea to go along with the phone: an online app store. The leadership team—the *same* widely admired, cover-story leadership team—shot down both projects. Three years later, the engineers saw their crazy ideas materialize on a stage in San Francisco. Steve Jobs unveiled the iPhone. Five years later, Nokia was irrelevant. It sold its mobile business in 2013. Between its mobile peak and exit, Nokia's value dropped by roughly a quarter *trillion* dollars.

A wildly innovative team had turned.

In medical research, for decades, Merck was the most revered company. From 1987 to 1993, it placed first in *Fortune*'s annual most-admired-company survey, a seven-year streak not matched until Apple achieved it in 2014. Merck launched the first cholesterol-lowering drug. It developed the first drug for river blindness, and then donated that drug at no cost to many countries in Africa and Latin America. Over the next ten years, however, Merck missed nearly every important breakthrough in drug discovery. It overlooked not only genetically engineered drugs, which transformed the industry (more about that later), but also drugs for cancer, autoimmune diseases, and mental disorders, the three biggest success stories of the 1990s and early 2000s.

In every creative field, we see legendary teams suddenly, and mysteriously, turn. In his wonderful memoir of his time at Pixar, Ed Catmull writes about Disney:

> After *The Lion King* was released in 1994, eventually grossing $952 million worldwide, the studio began its slow decline. It was hard, at first, to deduce why—there had been some leadership changes, yet the bulk of the people were still there, and they still had the talent and the desire to do great work.
>
> Nevertheless, the drought that was beginning then would last for the next sixteen years: From 1994 to 2010, not a single Disney animated film would open at number one at the box office. . . . I felt an urgency to understand the hidden factors that were behind it.

Let's talk about those hidden factors.

MORE IS DIFFERENT

The pattern of sudden changes in the behavior of teams and companies—of the same people suddenly behaving in very different ways—is a mystery in business and social science. Entrepreneurs, for example, often say that big companies fail because big-corporate types are conservative and risk-averse. The most exciting ideas come from small companies, because—we tell ourselves—we are the truly passionate risk-takers. But put that big-corporate type in a startup, and the tie will come off and he'll be pounding the table supporting some wild idea. The *same* person can act like a project-killing conservative in one context and a flag-waving entrepreneur in another.

The change in behavior may be a mystery in business, but a similar pattern is the essence of a strange quirk of matter called a *phase transition*. Imagine a large bathtub filled with water. Hit the surface with a hammer: a splash, and the hammer slips through the liquid. Then lower the temperature until the water freezes. Strike again, and the surface shatters.

The *same* molecule behaves like a liquid in one context and a rigid solid in another.

Why? How do molecules "know" to suddenly change their behavior? To put it another way, which brings us even closer to the mystery of our supposedly risk-averse, big-corporate type: If we drop a molecule of water onto a block of ice, what happens? It freezes. If we drop that same molecule into a pool of water, what happens? It slushes around with all the other molecules. How can we explain this?

The physicist and Nobel laureate Phil Anderson once captured the core idea underlying the answers to these questions with the phrase *more is different*: "The whole becomes not only more than but very different from the sum of its parts." He was describing not only the flow of liquids and

the rigidity of solids but even more exotic behaviors of electrons in metals (for which he won his Nobel Prize). There's no way to analyze just one molecule of water, or one electron in a metal, and explain any of these collective behaviors. The behaviors are something new: phases of matter.

I will show you that the same holds true for teams and companies. There's no way to analyze the behavior of any individual and explain the group. Being good at nurturing loonshots is a phase of human organization, in the same way that being liquid is a phase of matter. Being good at developing franchises (like movie sequels) is a *different* phase of organization, in the same way that being solid is a different phase of matter.

When we understand those phases of organization, we will begin to understand not only *why* teams suddenly turn, but also how to *control* that transition, just as temperature controls the freezing of water.

The basic idea is simple. Everything you need to know is in that bathtub.

WHEN SYSTEMS SNAP

The molecules of a liquid roam all over. Think of the water molecules in the tub as a platoon of cadets running randomly around a practice field. When the temperature drops below freezing, it's as if a drill sergeant blew a whistle and the cadets suddenly snapped into formation. The rigid order of the solid repels the hammer. The chaotic disorder of the liquid lets it slip through.

Systems snap when the tide turns in a microscopic tug-of-war. Binding forces try to lock water molecules into rigid formation. Entropy, the tendency of systems to become more disordered, encourages those molecules to roam. As temperature decreases, binding forces get stronger and entropy forces get weaker.

When the strengths of those two forces cross, the system snaps. Water freezes.

All phase transitions are the result of two competing forces, like the tug-of-war between binding and entropy in water. When people organize into a team, a company, or any kind of group with a mission they also create two competing forces—two forms of incentives. We can think of the two competing incentives, loosely, as *stake* and *rank*.

When groups are small, for example, everyone's *stake* in the outcome of the group project is high. At a small biotech, if the drug works, everyone will be a hero and a millionaire. If it fails, everyone will be looking for a job. The perks of *rank*—job titles or the increase in salary from being promoted—are small compared to those high stakes.

As teams and companies grow larger, the stakes in outcome decrease while the perks of rank increase. When the two cross, the system snaps. Incentives begin encouraging behavior no one wants. Those same groups—with the same people—begin rejecting loonshots.

The bad news is that phase transitions are inevitable. All liquids freeze. The good news is that understanding the forces allows us to manage the transition. Water freezes at 32 degrees Fahrenheit. On snowy days, we toss salt on our sidewalks to lower the temperature at which water freezes. We want the snow to melt rather than harden into ice. We'd rather wet our shoe in a puddle than slip and spend a week in the hospital.

We use the same principle to engineer better materials. Adding a small amount of carbon to iron creates a much stronger material: steel. Adding nickel to steel creates some of the strongest alloys we know: the steels used inside jet engines and nuclear reactors.

We will see how to apply a similar principle to engineer more innovative organizations. We will identify the small changes in *structure*, rather than *culture*, that can transform a rigid team.

Leaders spend so much time preaching innovation. But one desperate molecule can't prevent ice from crystallizing around it as the temperature drops. Small changes in structure, however, can melt steel.

* * *

This book is divided into three parts. Part one tells five stories of five remarkable lives. The stories illustrate a central idea: why being good at loonshots (like original films) and being good at franchises (sequels) are phases of large-group behavior—distinct and separate phases. No group can do both at the same time, because no system can be in two phases at the same time. But there's one exception. When the water in the bathtub mentioned earlier is at exactly 32 degrees Fahrenheit, pockets of ice coexist with pools of liquid. Just below or above that temperature, the whole thing will freeze or liquefy. But right at the edge of a phase transition, two phases can coexist.

The first two rules for nurturing loonshots, described in part one, are the two principles that govern life on the edge. A third rule explains how to hold that edge long-term. It borrows from chess rather than physics: the longest-reigning chess champion in history ascribed much of his success to mastering this idea.

Part two describes the underlying science. We'll see how the science of phase transitions has helped us understand the spread of wildfires, improve traffic flow, and hunt terrorists online. We'll apply similar ideas to see why teams, companies, or any group with a mission will snap between two phases just like the water in the bathtub snaps between liquid and solid.

Putting these pieces together will reveal the science behind the "magic number 150": an equation that describes when teams and companies will turn. That equation will lead us to an additional rule that shows us how to *raise* the magic number—a change that will make any loonshot group more powerful. (The four rules, as well as four more personal lessons for anyone nurturing any kind of loonshot, are summarized at the end.)

A final chapter describes what we might call the mother of all loonshots. We'll extend these ideas on the behavior of groups to the behavior of societies and nations, and see how that helps us understand the course of history: why tiny Britain, for example, toppled the far larger and wealthier empires of India and China.

This may all sound a bit . . . loony.

That's the idea.

* * *

To begin, we will turn to an engineer handed a national crisis.

Let's turn to the brink of world war.

PART ONE

ENGINEERS OF SERENDIPITY

1

How Loonshots Won a War

Life on the edge

Had there been prediction markets in 1939, the odds would have favored Nazi Germany.

In the looming battle between world powers, the Allies lagged far behind Germany in what Winston Churchill described as the "secret war": the race for more powerful technologies. Germany's new submarines, called U-boats, threatened to dominate the Atlantic and strangle supply lines to Europe. The planes of the Luftwaffe, ready to bomb Europe into submission, outclassed those of any other air force. And the discovery of nuclear fission early that year, by two German scientists, put Hitler within reach of a weapon with almost unfathomable power.

Had the technology race been lost, Churchill wrote, "all the bravery and sacrifices of the people would have been in vain."

By the time Vannevar Bush, dean of engineering at MIT, quit his job, moved to Washington, and talked his way into a meeting with the president in the summer of 1940, the US Navy already held the key to winning that race. They'd had it for eighteen years. They just didn't know it.

To find that key and win that race, Bush invented a new system for nurturing radical breakthroughs.

It was the secret recipe for winning the secret war.

THE *DORCHESTER*

In late September 1922, two ham-radio enthusiasts at the US Naval Air Station just outside Washington, DC, set up a shortwave radio transmitter on the edge of the station overlooking the Potomac River. Leo Young, 31, from a small farm town in Ohio, had been building radio sets since high school. His partner, Hoyt Taylor, 42, a former physics professor, was the Navy's senior radio scientist. They'd come together to test whether high-frequency radio could help ships communicate more reliably at sea.

Young rigged the radio's transmitter to operate at 60 megahertz, 20 times higher than the level for which it had been designed. He jacked up the sensitivity of its receiver using a technique he'd discovered in an engineering journal. Equipment suitably tweaked, the two turned on the transmitter, loaded the receiver onto a truck, and drove to Hains Point, a park directly across the Potomac from the naval air station.

They placed the receiver on the stone seawall at the edge of the park and aimed it at the transmitter across the river. The receiver emitted the steady tone of a clear signal. Suddenly, the tone doubled in volume. Then it disappeared completely for a few seconds. Then it came back at double volume for a moment before settling back to the original, steady tone.

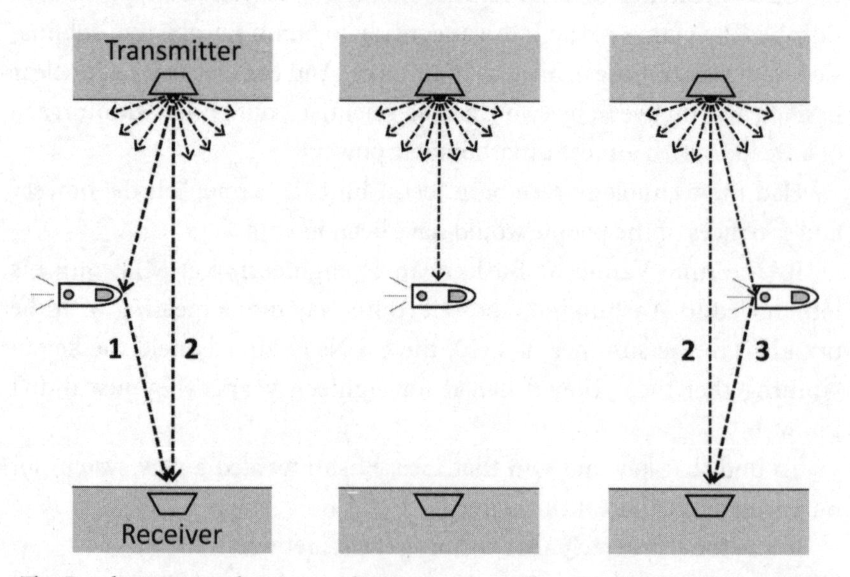

The *Dorchester* passes between radio transmitter and receiver in the Potomac River

They looked up and saw that a ship, the *Dorchester*, had passed between the receiver and the transmitter.

To the two engineers, the doubling in strength was an unmistakable sign of something called radio-wave interference: two synchronized beams adding together. When the hull of the *Dorchester* reached a "sweet-spot" distance from the line of sight between transmitter and receiver, the beam bouncing off the hull (beam #1 on the left in the figure on the previous page) traveled through a path exactly one-half of a radio wavelength longer than the line-of-sight beam (beam #2). At that point, the two beams precisely synchronized, which explained why the tone from the receiver doubled. As the ship passed through the line of sight, it blocked the signal completely. After the ship cleared the line of the sight, on the right in the figure, the tone came back. When the back of the boat reached the same sweet-spot distance from the line of sight, the reflected and direct beams synchronized precisely again. That explained the second doubling in tone.

Young and Taylor were testing a tool for communication. But they had accidentally discovered a tool for detection.

The two engineers repeated the experiment successfully several more times, and a few days later, on September 27, they sent a letter to their superiors describing a new way to detect enemy ships. A line of US ships carrying receivers and transmitters could immediately detect "the passage of an enemy vessel . . . irrespective of fog, darkness or smoke screen."

This was the earliest known proposal for the use of radar in battle. One military historian would later write that the technology changed the face of warfare "more than any single development since the airplane."

The Navy ignored it.

With no support for their proposal and their request for funding rejected, Young and Taylor abandoned the idea. They worked on other radio projects for the Navy—but they didn't forget. Eight years later, in early 1930, Young and another engineer at the lab, Lawrence Hyland, set about testing a new idea for guiding the landing of planes. A transmitter on the ground near a landing strip would beam a radio signal into the sky; the pilot in an approaching plane would direct his plane to follow the signal and land. One hot, muggy afternoon in June, in a field two miles from the upward-pointing transmitter, Hyland began testing the receiver they planned to use. As he adjusted the equipment, his receiver suddenly grew loud and noisy. Then it quieted down. A few moments later, it grew loud again. Then it settled down again. The pattern persisted. He checked and rechecked his equipment and couldn't find a problem. As he prepared to return his broken receiver to the lab, he noticed something odd: the signal got loud whenever a plane flew overhead.

Hyland told Young, who quickly realized the connection with what he had seen years earlier on the Potomac. The beam aimed into the sky bounced off an overhead plane and landed in Hyland's receiver. Reflecting radio waves, as they soon confirmed, could detect not only ships but planes flying as high as eight thousand feet, even when those planes were miles away. They conducted detailed tests and, once again, submitted a proposal for something never seen before in warfare: an early warning system for enemy aircraft.

Nothing happened. A request for $5,000 in funding was rejected because the time to see results "might well exceed two or three years." Another desk chief wrote dismissively that the idea was "a wild dream with practically no chance of real success," listing a handful of reasons it was impractical. It took five years for the military to assign one full-time person to the project.

One career officer who fought a mostly losing battle inside the Navy to accelerate development of radar recalled later, "It really pained me . . . to think how much two years of fleet experience with radar before 1941 could have saved us in lives, planes, ships and battles lost during the initial phases of the Pacific war."

A radar early warning system was still being field-tested in Hawaii on the morning of December 7, 1941.

The surprise attack on Pearl Harbor, by 353 enemy aircraft, killed 2,403 people.

HOW NOT TO FIGHT A WAR

Like Miller's piranha, described in the introduction, Young and Taylor's discovery was a classic loonshot. The idea that would turn the course of the war passed through a decade-long tunnel of neglect and skepticism.

Into that tunnel strode a man with an uncommon ability to see past common doubts—Vannevar Bush, a tall, thin, upstanding preacher's son, who swore like a sailor and dressed like a tailor. When the First World War began, Bush had just completed a graduate degree in engineering. He volunteered at the submarine research station in New London, Connecticut. His experience there would be similar to that of Young and Taylor eight years later. The Navy buried his most valuable idea: a magnetic device for detecting submerged submarines. From the experience, Bush wrote, he learned "how not to fight a war." In the high-stakes competition between weapons and counterweapons, the weak link was not the supply of new ideas. It was the transfer of those ideas to the field.

Transfer requires trust and respect on both sides. But officers "made it utterly clear that scientists or engineers employed in these laboratories were of a lower caste of society," Bush wrote, referring to New London and similar centers. At the start of that war—the first in which poison gas was used—the secretary of war rejected an offer of help from the American Chemical Society because he "had looked into the matter and found the War Department already had a chemist."

Despite that friction, Bush chose to maintain his ties with the Navy after the war. It forced him to learn a new skill: the ability to embrace others unlike himself, a skill that would later prove immensely valuable. Bush served in the naval reserves for eight years, even as his career as an academic, an engineer, and a businessman grew. He was appointed a professor of engineering at MIT, invented one of the earliest computers (an analog machine), and helped launch a company that grew into the massive electronics manufacturer Raytheon.

By the mid-1930s, Bush had risen to second in command to the president of MIT, and he was still consulting for the Navy. What he saw in the military alarmed him. Despite the growing threat from fascism in Europe and Asia, the armed services in 1936 cut funds for research on new technologies to one-twentieth the cost of one battleship. An Army memo explained that the only force that mattered was "the infantry with rifle and bayonet." Bush warned of a growing technology gap with Germany. But little had changed since his experience in New London. Generals had no interest in the views of "damn professors," their term for civilian scientists.

By 1938, Hitler had annexed Austria and the Sudetenland; Franco and his Nationalists had captured most of Spain; Mussolini was in full control of Italy; and Japan had invaded China and captured Beijing (then called Peking). Bush and a handful of other scientific leaders—including James Conant, a chemist and the president of Harvard University—believed war was coming and the US was dangerously unprepared. Both had witnessed the tendency of generals to fight a war with the weapons and tactics of the preceding war. They understood that the same mistake this time—facing a much greater German threat—could be fatal.

The military, Bush knew, had been gearing up to produce more of the same: more planes, more ships, more guns. Like a large film studio churning out sequel after sequel, the military was operating in what we might call a franchise phase.* To invent the radically new technologies necessary to defeat the Germans, however, the military would need to operate in an entirely different phase, one that offered scientists and engineers, as Bush wrote, the "independence and opportunity to explore the bizarre."

In other words, Bush understood intuitively that being good at franchises and being good at loonshots are phases of organization. And the same organization can't be in two phases at the same time, for the same reason water can't be both solid and liquid at the same time—under ordinary conditions.

Ordinary conditions did not apply in 1938. The generals really would need munitions built at an unprecedented rate, troops and supplies distributed across four continents, and millions of soldiers directed in battle. But the military would also need to win Churchill's secret war: the race to create technologies that did not yet exist.

To survive, the country needed both.

One molecule can't transform solid ice into liquid water by yelling at its neighbors to loosen up a little. Which is why Bush didn't try to change military *culture*. A different kind of pressure is required. So Bush created a new *structure*. He adopted the principles of life on the edge of a phase transition: the unique conditions under which two phases can coexist.

In April 1944, a glowing *Time* profile would describe Vannevar Bush as the general of a secret army of scientists that "is regarded almost with awe" in Washington. In October 1945, the Committee on Appropriations of the US House of Representatives would declare that without Bush's organization, "it is safe to say that victory still would await achievement."

But in 1938, Bush's battles were just beginning.

---

* The term "franchise" is a convenient shorthand used in film and drug discovery and certain other businesses. The reason to use the term will become clearer later.

GATHERING STORM

By the mid-1930s, Bush had become widely known for his skill in bridging science, industry, and government. So it came as no surprise when, in 1938, the Carnegie Institution, a Washington-based think tank that supports scientific research, offered Bush their top job. In response, the president of MIT offered to step down and make Bush president of the university if he would stay.

Bush declined. Although a prestigious career and generations of New England family ties rooted him in Boston, Bush understood that the nation's defense would be led from Washington. And no one else had his ability to bridge worlds. He was uniquely qualified, he knew, to mobilize the nation's scientists for war.

"All of [my] recent ancestors were sea captains, and they have a way of running things without any doubt," Bush said years later. "So it may have been partly that, and partly my association with my grandfather, who was a whaling skipper, [which] left me with some inclination to run a show, once I was in it."

Bush quit his job, accepted the Carnegie offer, and moved to Washington.

With the help of the Carnegie trustees, one of whom was President Franklin Roosevelt's uncle, Bush put together a plan. "I knew that you couldn't get anything done in that damn town," he recalled, "unless you organized under the wing of the President."

A place for Bush under that wing seemed unlikely. The president, a lawyer surrounded by social planners, had shown little interest in science or scientists. Bush, a conservative by nature and upbringing, was in turn skeptical of both FDR and his New Deal aides. He'd grown up distrusting "social innovators," whom he thought of "as a bunch of long-haired idealists or do-gooders."

Bush leaned on the president's uncle to get a meeting with Harry Hopkins, Roosevelt's closest advisor. Hopkins, a former social worker and a do-gooder of the highest order, was an equally unlikely ally. Years later Bush wrote, "The fact that Harry and I hit it off is among the minor miracles." But hit it off they did—Hopkins had a taste for bold ideas.

On June 12, 1940, at 4:30 p.m., Bush and Hopkins met with Roosevelt in the Oval Office. Their message: the Army and Navy trailed far behind

Germany in the technologies that would be critical to winning the coming war. The military on its own was incapable of catching up in time. Bush proposed that FDR authorize a new science and technology group within the federal government, to be led by Bush, reporting directly to the president.

FDR listened, read Bush's proposal—four short paragraphs in the middle of one piece of paper—and signed it "OK—FDR." The meeting lasted all of ten minutes.

Bush's new organization, eventually called the Office of Scientific Research and Development (OSRD), would create the opportunity Bush sought for scientists, engineers, and inventors at universities and private labs to explore the bizarre. It would be a national department of loonshots, seeding and sheltering promising but fragile ideas across the country. The group would develop the unproven technologies the military was unwilling to fund. It would be led by a *damn professor.*

The military and its supporters, as expected, objected. They told Bush his new group "was an end run, a grab by which a small company of scientists and engineers, acting outside established channels, got hold of the authority and money for the program of developing new weapons."

Bush's answer: "That, in fact, is exactly what it was."

LIFE AT 32 FAHRENHEIT

Imagine bringing that bathtub to the brink of freezing. A little bit one way or the other and the whole thing freezes or liquefies. But right on the cusp, blocks of ice coexist with pockets of liquid. The coexistence of two phases, on the edge of a phase transition, is called *phase separation*. The phases break apart—but stay connected.

The connection between the two phases takes the form of a balanced cycling back and forth: Molecules in ice patches melt into adjacent pools of liquid. Molecules of liquid swimming by an ice patch lock onto a surface and freeze. That cycling, in which neither phase overwhelms the other, is called *dynamic equilibrium*.

As we will see, *phase separation* and *dynamic equilibrium* were the key ingredients in Bush's recipe. "The essence of a sound military organization is that it should be tight. But a tight organization does not lend

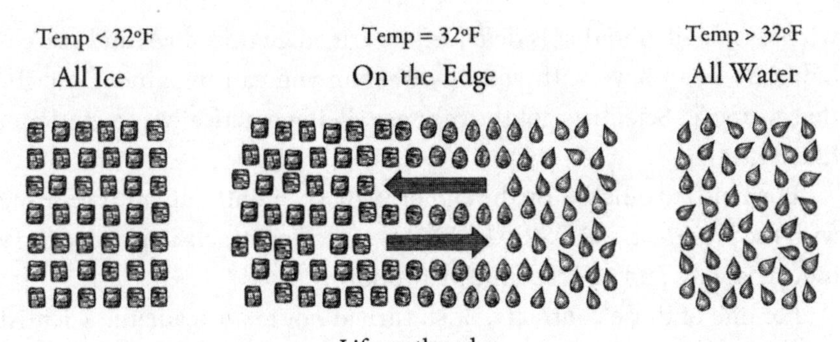

Life on the edge

itself to innovations," Bush wrote. "And loosening it in time of war . . . would be fraught with danger." But, Bush continued, there "should be close collaboration between the military and [some] organization, made loose in its structure on purpose."

In other words, the two phases must *break apart* while *staying connected.*

Bush's attempt to apply the first of these two principles, phase separation—a new agency entirely under his control—did not begin well. One officer explained to Bush "that no damned civilian could possibly understand a military problem." Bush's reaction: "I waded into him . . . and said that, unfortunately, there were still some officers in existence who were so dense that they did not realize that the art of war was being revolutionized all about them."

Another high-ranking officer, on seeing the proposal from Bush's group for a new kind of amphibious truck, told Bush "that the Army did not want it and would not use it if they got it." (Bush ignored him. The truck, called the DUKW, would be widely used in the second half of the war.) Bush's former colleagues, university scientists, were equally skeptical of creating any ties to the military. They interpreted any federal oversight as interference.

Bush brought the two groups together. He used his credibility as an academic to reassure scientists of their independence. But at the same time, he explained their goal was more than clever ideas. Their goal was products that worked. When interviewing new scientists for his team, he would pose a challenge: "You are about to land at dead of night in a rubber raft on a German-held coast. Your mission is to destroy a vital enemy wireless installation that is defended by armed guards, dogs, and searchlights. You can have with you any weapon you can imagine. Describe that weapon." Scientists got the message. Being practical was a matter of life or death.

Bush moved quickly. By the end of 1940, six months after his meeting with the president, the OSRD had 126 research contracts in place with 19 industrial labs and 32 academic institutions.

For one of those contracts, Bush turned not to an academic scientist or to an industrial lab but to a wealthy investment banker named Alfred Lee Loomis, an expert in chess and magic tricks, who wore perfectly pressed white suits and lived a double life. By day he worked on Wall Street. On evenings and weekends, he retired to a massive stone castle forty miles away in Tuxedo Park, New York. The castle was a semisecret, private research lab, brimming with equipment built or purchased to satisfy the curiosity of its owner. In the mid-1930s, guests visiting Loomis's castle might be guided into a comfortable chair as an assistant materialized with small scissors, snipped some hair, swabbed alcohol onto their scalp, affixed electrodes, and encouraged them to relax. They had just become subjects of his research. (Loomis was an early pioneer of electroencephalography—EEG.)

From Albert Einstein, Enrico Fermi, and other European scientists who visited his lab, Loomis learned disturbing news of advanced German science applied to weapons of war as well as hints of a terrifying German discovery in nuclear physics. Like Bush and Conant, Loomis had worked with the US military during the First World War. Also like them, he had concluded that the Army and Navy were incapable of catching up to the German lead on their own. So when Loomis got the call to join Bush's new organization, he dropped all other projects. At the suggestion of Bush and his advisors, Loomis began working full-time on a new technology: microwave radar.

By the end of 1940, Loomis had assembled dozens of the country's best engineers and physicists inside an anonymous building at MIT. Their goal was to develop a radar system using short wavelengths (ten centimeters, called microwaves) rather than long wavelengths (tens or hundreds of meters, called radio waves). The shorter the wavelength, the greater the resolution. The radio-wave system developed at the naval lab (and later discovered independently in Britain) could detect ships and planes. A microwave system could detect the periscope of a submarine or track an incoming missile. But an even more important advantage had to do with size. Wavelength determines the size of the antenna needed, which is why microwave ovens fit in your kitchen and radio towers do not. A microwave radar system, if they could build one, would be portable. Any ship, plane, or even truck could carry one.

While Loomis was beginning his work on radar, a team in Britain neared completion of a national radar defense system. (The British discovery of radar was due, in part, to public demands that their Air Ministry investigate the use of death-ray weapons. The most insistent requests came from a widely ignored former government official given to ranting about imagined future air attacks on London. His name was Winston Churchill.) By the late 1930s, a chain of radar antennas ringed the British coast.

After Germany marched through Poland in the fall of 1939 and made quick work of the rest of Europe in the spring of 1940, Hitler turned his attention north. In June, Churchill announced to Parliament, "The battle of Britain is about to begin . . . Hitler knows that he will have to break us in this island or lose the war."

Churchill continued with what became one of the most famous lines of the twentieth century: "Let us therefore brace ourselves to our duties, and so bear ourselves that, if the British Empire and its Commonwealth last for a thousand years, men will still say, 'This was their finest hour.'"

In July, Hitler attacked. His generals anticipated that the Luftwaffe, which had twice as many planes as the British Royal Air Force (RAF), would achieve air superiority in two to four weeks, as it had across continental Europe. Hitler's generals developed plans for a land invasion of Britain, called Operation Sea Lion, to follow the victory in the air.

That victory would never come. Britain's chain of radar antennas allowed the RAF to detect enemy aircraft before they neared the coast. The intelligence allowed the British to concentrate their limited forces against each wave of attack. On September 15, commemorated in England as Battle of Britain Day, 144 German pilots and crew were shot down versus only 13 for the RAF. One German bomber pilot, whose unit had lost one-third of its aircraft in one hour, wrote that "if there were any more missions like that, our chances of survival would be nil."

Two days later, Hitler indefinitely postponed the land invasion of Britain. By the end of October, the German attack was all but over. It was Germany's first loss of the war.

At the time, British-American relations were delicate. The Americans were still officially neutral, and FDR was under pressure from isolationists to stay out of the war. The American ambassador to London, Joseph Kennedy, had widely shared his views that England would quickly fold under a German attack (one British diplomat described Kennedy as a "foul specimen of double-crosser and defeatist"). And a clerk at the American embassy in London, who had full access to the most secret correspondence between Churchill and FDR, had been exposed as a German spy.

Yet on August 6, 1940, Churchill authorized a British scientific mission to the US. They were to reveal everything they knew about radar to Alfred Loomis and his team.

The technologies they shared jump-started Loomis's efforts. And the urgent need for something new soon became painfully clear.

MASSACRE

In February 1941, four months after the German defeat in the air battle over Britain, Hitler issued a new directive. If Germany couldn't bomb England into submission, it would starve her to death: a siege. The principal weapon of that siege would be the U-boat. Unfortunately for the Allies, the long-wavelength radar used in the Battle of Britain proved useless against submarines. Long-range antennas required too much energy and were too heavy to mount on ships or planes. Sonar also did little to stop Hitler's submarines: the range was too short, and sonar could not detect subs on the surface.

Allied losses to U-boats rose rapidly, from 750,000 tons of cargo in 1939 to 4.3 million tons in 1941. Every month, U-boats were sinking ships faster than the Allies could build them. And the losses kept mounting.

At the end of that year, on December 11, four days after Pearl Harbor, Hitler declared war on the United States. He unleashed Vice Admiral Karl Doenitz, his commander in charge of submarines, to fire at will on US ships in the Atlantic.

Unlike Great Britain, the US had no recent experience fighting submarines. Bright boardwalk lights from amusement parks and casinos spilled into the dark ocean night, skimming over the waves, guiding U-boat commanders to the coast. One German officer, stunned by the contrast with the pitch-dark blackouts enforced in Europe, wrote, "We were passing the silhouettes of ships recognizable in every detail . . . they were formally presented to us on a plate: please help yourselves!"

On January 13, Captain Reinhard Hardegen, commanding the long-distance, Type IX U-boat U-123, glided into New York Harbor. Shortly after midnight, he noticed a ship approaching, port side, its lights and lanterns glowing. He raised his field glasses. "It's a tanker," he told his watch officer. "A huge one." He swung south to head at a right angle to the ship's path. At 800 meters, he ordered two torpedoes. For a silent minute, they streaked through the water. Then the blast wave of an explosion rocked his sub. Flames from the tanker shot into the sky and turned into a "black, sinister mushroom cloud 150 meters high." The *Norness*, a 9,577-ton tanker, became the first casualty in a wave of destruction off the US coast, in which a handful of U-boats destroyed or damaged nearly 400 ships and killed nearly 5,000 passengers and crew.

In his war memoirs, Churchill described the Allies' ability to protect their fleets as "hopelessly inadequate . . . week by week the scale of this massacre grew."

Allied shipping losses reached a staggering 7.8 million tons in 1942. By early 1943, food supplies to Britain had dwindled to two-thirds of normal levels. The government was forced to ration basic goods. Less than three months of commercial oil reserves remained—ten months, if all emergency military reserves were included. Oil supply was not publicly discussed, but every commander, on both sides of the Atlantic, paid close attention. No oil meant no planes, no ships, no transport. No ability to resist the German machine. And Britain was running on fumes.

In early March 1943, German codebreakers deciphered Allied transmissions indicating two large convoys, over 100 ships in total, heading eastward. Forty-three U-boats rushed to intercept them. Within 48 hours, the U-boats sank 20 ships without suffering a single loss.

The British *Canadian Star* was hit on March 18. One survivor recalled the scene: "The sea just swept through the boat from end to end. I could see the men, one by one, their eyes glazing and eventually losing their grip and being washed up and down the boat and eventually out [to sea]."

The ship's carpenter, 58 years old, decided he had no chance. "He

The Battle of the Atlantic

called out to one of the ship's officers, 'Goodbye, Sir. It was a good life while it lasted,' waved and then calmly 'walked right into the path of a wave pounding across the afterdeck. It was like a minnow being swallowed by a whale.'"

In Berlin, Doenitz and his staff celebrated: they had inflicted the largest single maritime loss of the war.

It would be their last celebration.

The same month the *Canadian Star* went down, US Army Air Force B-24 Liberator bombers, equipped with two new devices created by Loomis and his team of loons, deployed over the Atlantic. The first device was a powerful microwave radar. Developed in less than 30 months, it could detect the periscopes of surfaced submarines, day or night, through clouds or fog.

Hunting subs across the vast ocean, however, required planes to quickly locate and fly to convoys when summoned. Navigating by the stars would be impossible, especially in rough weather. Loomis and his team came up with another idea: a grid of pulsed radio signals covering the Atlantic. With a specially designed decoder, a pilot could calculate his location on the grid without alerting an enemy ship.

By the spring of 1943, the long-range Liberator bombers, with microwave radar and pulsed-radio navigation, were fully operational and patrolling the Atlantic.

ONE AT A TIME

On May 11, a convoy of 37 ships designated SC-130 departed Canada sailing eastward for England. Six days later, German intelligence identified the route through intercepted signals and alerted a wolf pack of 25 subs. On the evening of May 18, deep in the mid-Atlantic, the convoy encountered the first U-boats from the pack. The commander of the convoy's escort ships, Peter Gretton, radioed for backup. Liberator bombers arrived within hours. The bombers' microwave radar cut through darkness and fog. Previously invisible subs lit up their oscilloscope screens.

Gretton and the bombers hunted every U-boat that showed itself. To escape the depth charges and guns, U-boats dove deep the instant they saw a plane or destroyer. *U-645* radioed back to Berlin: "UP TO NOW HAVE BEEN DRIVEN UNDER WATER CONTINUALLY BY AIRCRAFT OUT OF LOW CLOUDS AND BY DESTROYERS." *U-707*: "CONTINUOUSLY DRIVEN UNDER WATER." When Liberator P/120 arrived and quickly spotted a handful of subs, the pilot radioed Gretton for target priorities. Gretton replied with a list. The pilot joked, "As Mae West said, one at a time, gentlemen, please."

During the three-day battle, the German U-boats were unable to launch a single successful attack. Back in Berlin, Doenitz received similar messages from U-boat commanders all across the Atlantic: they were being continuously chased underwater by bombers, with losses mounting.

They had gone from the hunters to the hunted.

On May 20, Doenitz radioed the pack of U-boats fighting convoy SC-130: "BREAK OFF OPERATIONS AGAINST CONVOY." The battle was over. Not one Allied ship was lost. Three U-boats were sunk, with all aboard lost at sea, including one sub that contained a 21-year-old officer on his first mission: Doenitz's son.

In total, Allied planes and ships sank 41 U-boats in May, more in one *month* than in any of the first three years of the war. The loss came to nearly one-third of Doenitz's total operational fleet. On May 24, recognizing the inevitable, Doenitz withdrew U-boats from the Atlantic. Later that year he wrote, "For some months past, the enemy has rendered the U-boat war ineffective. He has achieved this object, not through superior tactics or strategy, but through his superiority in the field of science; this finds its expression in the modern battle weapon—detection."

In 90 days, Allied shipping losses decreased by 95 percent: from 514,000 tons in March to 22,000 tons in June. "We had lost the Battle of the Atlantic," Doenitz wrote.

The U-boats never again threatened passage of a convoy. The lanes were cleared for an Allied invasion of Europe.

* * *

Radar had a far greater impact on the course of the war than is usually appreciated, extending well beyond the battle with the U-boats. Radar sighting from planes allowed the Allies to destroy enemy supplies, bridges, and transport with targeted bombing raids day and night, regardless of weather. Radar-controlled antiaircraft guns were essential to defending aircraft carriers, which created a decisive advantage in the Pacific War.

In June 1944, Germany unleashed upon London its V-1 "buzz bombs"—the first rocket-powered missiles, instantly recognizable by the terrifying insect-like buzzing sound that victims would hear as they approached. Developed at great expense and heralded by Hitler as a wonder weapon that could not be touched by enemy aircraft and would decimate the Allies, it was Germany's last-hope attack by air. Radar-sighting on guns allowed the missiles to be tracked and quickly shot down.

Radar also played a decisive role in the Battle of the Bulge in Belgium in late 1944—Germany's last-hope offensive on land, which caught the Allies by surprise. The Army used artillery shells with new, radar-equipped fuses. The fuses, designed to explode when a shell neared its target, improved firing efficiency by up to seven times (the equivalent of firing seven times as many guns). Later, after the Allied victory, General Patton said, "That funny fuse won the Battle of the Bulge for us."

* * *

Bush's system's ability to nurture loonshots with remarkable speed and efficiency was not limited to radar. The OSRD's

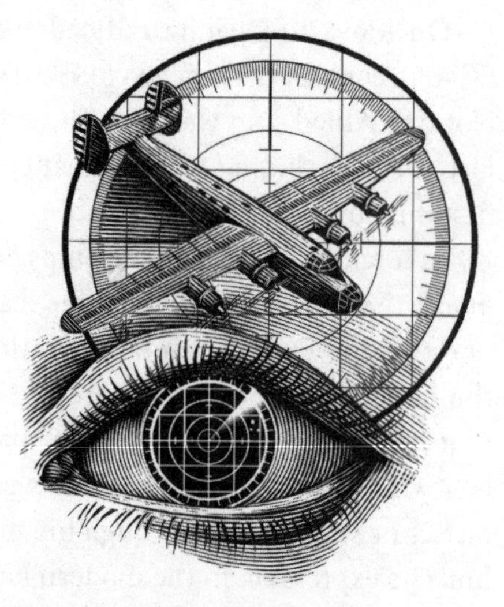

"Through superiority in the field of science."

work on penicillin, malaria, and tetanus contributed to lowering deaths from infectious disease among soldiers by a factor of twenty compared to the First World War. The work of OSRD scientists on plasma transfusion saved thousands of battlefield lives and grew into a standard hospital procedure after the war.

One invention, however—greeted with wonder at the time and horror soon after—overshadowed all others.

For the first two years after the 1939 discovery of nuclear fission—when the nucleus of an atom splits in two—most physicists believed the discovery had no practical applications, military or otherwise. A scientific committee appointed by President Roosevelt, after he received Einstein's famous letter warning of the threat from a new kind of bomb, concluded as much.

In 1941, new results from a group of atomic scientists in England convinced Bush otherwise. Bush made the case to FDR and Henry Stimson, his secretary of war, that even if the chances that a nuclear weapon could be built were small, the US could not risk Germany or Japan getting it first. FDR accepted his argument and put Bush in charge. Bush launched an aggressive research program, built support among military and political leaders, and then handed the program over to the military as the Manhattan Project.

The atomic bomb, when it arrived three years later, did not contribute to winning the war in Europe. Its role in ending the Pacific War is still controversial eight decades later. But had the US lost that race—and there was no way to know in advance whether the Axis powers would get there first—there is no doubt the world would be a much darker place today.

ENDLESS FRONTIER

In November 1944, after victory over Germany had become increasingly certain, FDR called Bush into his office.

> FDR: "What's going to happen to science after the war?"
> Bush: "It's going to fall flat on its face."
> FDR: "What are we going to do about it?"
> Bush: "We better do something about it quick."

Bush knew that US science had been poorly supported before the war, and that the nation's future well-being depended on reversing the country's dependence on others for basic research. "We can no longer count on ravaged Europe as a source of fundamental knowledge," he wrote later.

Shortly after that conversation, FDR sent Bush an official letter requesting that he outline a national plan for supporting science. FDR wrote that there was no reason the system Bush had created during the war "cannot be profitably employed in times of peace."

Although Bush didn't know it, FDR was suffering from severe cardiac disease and possibly metastatic cancer. In his letter, FDR emphasized medical research:

> The fact that the annual deaths in this country from one or two diseases alone are far in excess of the total number of lives lost by us in battle during this war should make us conscious of the duty we owe future generations. . . .
>
> New frontiers of the mind are before us, and if they are pioneered with the same vision, boldness, and drive with which we have waged this war we can create a fuller and more fruitful employment and a fuller and more fruitful life.

Bush's report, called *Science: The Endless Frontier*, presented to President Truman in June 1945, two months after FDR's death, and released the following month, caused a sensation. The country had no national science policy, he declared. Philanthropy and private industry could not be relied upon to fund the basic research that is "the pacemaker of technological progress," essential for national security, economic growth, and the fight against disease. The report outlined the architecture of a new national research system.

Within days of its publication, Bush's report was hailed across the major news outlets. The *New York Times*, however, questioned its conclusions and patiently explained the nature of science to Bush (and his 41 MD and PhD coauthors): "The scientific method is always the same, whether it deals with radar or disease. Dr. Bush's report ignores this fact." The *Times* concluded by suggesting a better model: "Soviet Russia has approached this task more realistically."

BusinessWeek, in any event, which approvingly described Bush as "a practical businessman as well as a scholar," stated that *Endless Frontier* was "epoch-making" and "must-reading for American businessmen."

BusinessWeek turned out to be more prophetic than the *New York Times*. Since the end of World War II, hundreds of industry-changing, or industry-creating, discoveries originating in the US—including GPS, personal computers, the biotechnology industry, the internet, pacemakers, artificial hearts, magnetic resonance imaging, the chemotherapy cure for childhood leukemia, even the original Google search algorithm—sprang from the system Bush's report inspired. Many others were the joint offspring of public and private research. Without the federal investment in the band theory of solids, for example, and techniques for growing high-quality germanium and silicon crystals, there would have been no transistor to launch the electronics age (more on the transistor later).

Quantifying the impact of these discoveries, and separating the contribution of private vs. public investment, is difficult. But as one measure, economists have attributed roughly *half* of the trillions of dollars in US GDP growth since the end of World War II to technology improvements.

Although neither Bush nor FDR could have foreseen the growth that would be created from "profitably employing" Bush's ideas in times of peace, both did have practical business experience. Bush's system, in fact, came from the business world.

EIGHT NOBEL PRIZES

Vannevar Bush rescued a large organization, dominated by a powerful franchise, in a crisis caused by a failure to innovate: the US military. In 1907, another large organization dominated by a powerful franchise had been deep in crisis for the same reason.

Thirty years after Alexander Graham Bell and his father-in-law created the Bell Telephone Company, its continued survival was in serious doubt. The company faced rapidly deteriorating finances, growing competition from the thousands of new phone companies created after Bell's telephone patent expired, and a public angry about the phone system's declining quality. The company's leaders had been milking the license from Bell's patent, collecting their checks, and not doing much else.

In 1907, a banking group led by the financier J. P. Morgan took control of the company, by then renamed AT&T, and got rid of its management. Morgan installed Theodore Vail, age 62, as its new chief executive.

Shortly after Vail took over, he promised that Americans would soon be able to call anyone, anywhere in the country, from New York to San Francisco. Few inside AT&T believed Vail. Calls over even a fraction of that distance barely worked. Electric signals faded as they traveled down a wire, and no one could explain exactly why. The electron had only been discovered ten years earlier. Quantum mechanics, which held the answer, was twenty years away. Vail's goal required technologies that did not yet exist, based on science that was not yet known.

Vail persuaded his new board of directors that to solve these problems, the company should create a quarantined group working on "fundamental" research. Like Bush, he understood the need for separating and sheltering radical ideas—the need for a department of loonshots run by loons, free to explore the bizarre. Vail put a physicist from MIT, Frank Jewett, in charge. Over the next several years, Jewett's group worked through the science and eventually solved the problem of the fading signals. They invented the vacuum tube: the world's first amplifier, the forerunner of all modern electronics.

Less than eight years after Vail took over, on January 25, 1915, several hundred people gathered in his fifteenth-floor conference room in New York. Alexander Graham Bell, coaxed out of retirement, placed a call to Thomas Watson in San Francisco. It was 39 years after their first wired conversation between two buildings in Boston. Watson picked up.

"Mr. Watson, come here, I want you," said Bell.

"It would take me a week to get to you this time," replied Watson.

Over the next 50 years, Vail's organization—eventually called the Bell Telephone Laboratories—produced the transistor, the solar cell, the CCD chip (used inside every digital camera), the first continuously operating laser, the Unix operating system, the C programming language, and eight Nobel Prizes. Vail created the most successful industrial research lab in history, and AT&T grew into the country's largest business.

Vail's protégé and the president of Bell Labs, Frank Jewett, would first meet Bush when they both worked on submarine detection for the Navy during World War I. Over the subsequent three decades, Jewett became a close friend and mentor to Bush. During World War II, Bush picked Jewett to serve as one of the five core members of his team. Many of the principles that Bush applied during the war had first been applied by Vail and Jewett at Bell Labs.

Bush changed national research the same way Vail changed corporate research. Both recognized that the big ideas—the breakthroughs that change the course of science, business, and history—fail many times before they succeed. Sometimes they survive through the force of exceptional skill and personality. Sometimes they survive through sheer chance. In other words, the breakthroughs that change our world are born from the marriage of genius and serendipity.

The magic of Bush and Vail was in engineering the forces of genius and serendipity to work for them rather than against them. Luck is the residue of design.

Now let's look a little more closely at that magic.

Vannevar Bush, James Conant, Karl Compton,
and Alfred Loomis at UC Berkeley (1940)

§

THE BUSH-VAIL RULES

There is a pervasive myth of the genius-entrepreneur who builds a long-lasting empire on the back of his ideas and inventions. (We will explore this myth, and the trap it creates, over the next several chapters.) But the ones who truly succeed—the engineers of serendipity—play a more humble role. Rather than champion any individual loonshot, they create an outstanding *structure* for nurturing many loonshots. Rather than visionary innovators, they are careful gardeners. They ensure that both loonshots and franchises are tended well, that neither side dominates the other, and that each side nurtures and supports the other.

The structures that these gardeners create share a common set of principles. I'll call these principles the Bush-Vail rules.

The first two rules are the ones mentioned above, the key to life at 32 Fahrenheit: separate the phases (the groups working on loonshots and on franchises) and create dynamic equilibrium (ensure that projects and feedback travel easily between the two groups). Break apart while staying connected.

1. SEPARATE THE PHASES

Separate your artists and soldiers

People responsible for developing high-risk, early-stage ideas (call them "artists") need to be sheltered from the "soldiers" responsible for the already-successful, steady-growth part of an organization. Early-stage projects are fragile. "Although military officers became avid for a new development once it had thoroughly proved itself in the field," Bush wrote, they dismissed any weapon "in embryo"—as they did with radar, with the DUKW truck, and with nearly every early innovation, which almost always arrives covered in warts. Without a strong cocoon to protect those early-stage ideas, they will be shut down or buried, like Young and Taylor's early discovery of radar.

Leaders of powerful franchises across every industry routinely dismiss early-stage projects by picking at their warts (the incentives behind this will be discussed more in part two). The major pharma companies passed on the idea for treating cancer by blocking tumor blood supply: the blood vessels known to surround tumors were dismissed as irrelevant inflammation. Major film studios passed on the idea of a metrosexual British spy who saves the world (the warts included a script in which the chief villain was a monkey). They also passed on a script titled at one point *The*

Adventures of Luke Starkiller, featuring an incomprehensible plot and a lead character named Mace Windy, suggesting a gassy superhero.

As we will see later, both industries have evolved a *structure* to rescue and nurture loonshots, despite the dominance of the powerful Majors. Warts were removed. Blocking tumor blood supply—called anti-angiogenesis therapy—became one of the great cancer breakthroughs of the past two decades. The first such drug, Avastin, achieved $7 billion in annual sales. The two unlikely film projects grew into the two most successful movie franchises of all time: James Bond and *Star Wars*.

The goal of phase separation is to create a *loonshot nursery*. The nursery protects those embryonic projects. It allows caregivers to design a sheltered environment where those projects can grow, flourish, and shed their warts.

Tailor the tools to the phase

Just separating loonshot and franchise groups is not enough. It's easy to draw a box on an org chart and rent a new building. But the list of failed companies with shiny research labs is long. True phase separation requires custom homes to meet custom needs: separate systems tailored to the needs of each phase.

Bush quarantined the team working on radar in anonymous office buildings at MIT. He recognized that the tight organization needed by the military, mentioned earlier, is not conducive to scientists exploring the bizarre, just as "a good organization for a research laboratory would not work well for a combat regiment in the field."

Vail quarantined the team working on the technology for long-distance telephony in an office building in lower Manhattan. Like Bush, he tailored the systems. He "moved away from the rigid task allocation" of telephone operations and toward a similar loose-touch style.

Both Bush and Vail understood intuitively decades ago what is repeatedly being rediscovered today. Efficiency systems such as Six Sigma or Total Quality Management might help franchise projects, but they will suffocate artists. When 3M, for example, inventor of Post-it Notes and Scotch Tape, brought in a high priest of Six Sigma as a new CEO in 2000, innovation plunged. It didn't recover until well after he left and a new CEO dialed back the system. The new CEO described the efficiency system as a mistake: "You can't say . . . well, I'm getting behind on invention, so I'm going to schedule myself for three good ideas on Wednesday and two on Friday." Art Fry, the retired inventor of Post-it Notes, said that his idea would never have emerged under the new approach.

Which doesn't mean that efficiency systems have no place. Loose goals and dream sessions might help artists. But they will harm the coherence of an army.

2. DYNAMIC EQUILIBRIUM

Love your artists and soldiers equally

Maintaining balance so that neither phase overwhelms the other requires something that sounds soft and fuzzy but is very real and often overlooked. Artists working on loonshots and soldiers working on franchises have to feel equally loved.

After creating what eventually became Bell Labs, Vail wrote, "No division, department, branch or group can be either ignored or favored at the expense of the others without unbalancing the whole." The trap for most groups, however, is that soldiers naturally favor soldiers and artists naturally favor artists.

Equal-opportunity respect is a rare and valuable skill. Vannevar Bush, although a veteran academic at the start of the war, genuinely respected the military. "I have enjoyed associating with military men more than with any other group, scientists, businessmen, professors," he wrote years later. The deference with which Bush treated officers helped him understand, and ultimately influence, the military far more than the many scientists and engineers who had tried, and failed, before him.

The less-famous history of an ultra-famous icon captures one person's evolution toward this balance. During Steve Jobs's first stint at Apple, he called his loonshot group working on the Mac "pirates" or "artists" (he saw himself, of course, as the ultimate pirate-artist). Jobs dismissed the group working on the Apple II franchise as "regular Navy." The hostility he created between the two groups, by lionizing the artists and belittling the soldiers, was so great that the street between their two buildings was known as the DMZ—the demilitarized zone. The hostility undermined both products. Steve Wozniak, Apple's cofounder along with Jobs, who was working on the Apple II franchise, left, along with other critical employees; the Mac launch failed commercially; Apple faced severe financial pressure; Jobs was exiled; and John Sculley took over (eventually rescuing the Mac and restoring financial stability).

When Jobs returned twelve years later, he had learned to love his artists (Jony Ive) and soldiers (Tim Cook) equally.

Although equal-opportunity respect is a rare skill by nature, it can be nurtured with practice (more on this in chapter 5).

Manage the transfer, not the technology

Bush, although a brilliant inventor and engineer, pointedly stayed out of the details of any one loonshot. "I made no technical contribution whatever to the war effort," he wrote. "Not a single technical idea of mine ever amounted to shucks. At times I have been called an 'atomic scientist.' It would be fully as accurate to call me a child psychologist."

Vail similarly stayed out of the details of the technical program. Both Bush and Vail saw their jobs as *managing the touch and the balance between loonshots and franchises*—between scientists exploring the bizarre and soldiers assembling munitions; between the blue-sky research of Bell Labs and the daily grind of telephone operations. Rather than dive deep into one or the other, they focused on the transfer between the two.

When the balance broke down, they intervened. As mentioned earlier, in the chain of creating a breakthrough, the transfer between the two sides is the weakest link. Scientists may pay little attention to soldiers or marketers. Soldiers and suits may dismiss the babble of nerds. Bush and

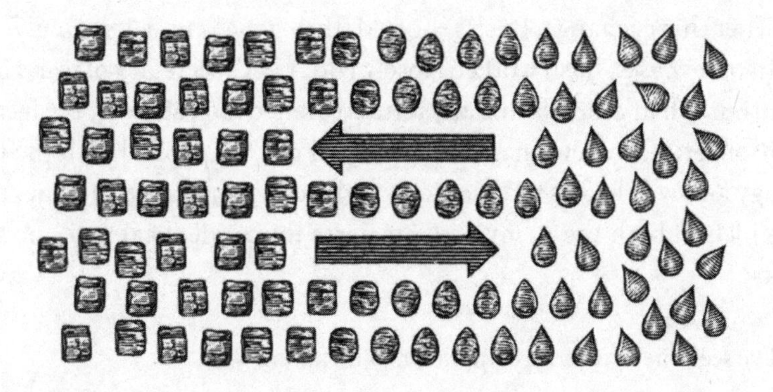

Vail zeroed in on that link. A radar detection device buried in a building full of physicists would sink no U-boats. A tiny switch made from semiconductors buried in Bell Labs would remain a curiosity rather than grow into the transistor, the invention of the century.

As we will see over the coming chapters, managing the touch and the balance is an art. Overmanaging the transfer causes one kind of trap. Undermanaging that transfer causes another.

A flawed transfer from inventors to the field is not the only danger. Transfer in the other direction is equally important. No product works perfectly the first time. If feedback from the field is ignored by inventors, initial enthusiasm can rapidly fade, and a promising program will be dropped. Early aircraft radar, for example, was practically useless; pilots ignored it. Bush made sure that pilots went back to the scientists and explained *why* they weren't using it. The reason had nothing to do with the technology: pilots in the heat of battle didn't have time to fiddle with the complicated switches on the early radar boxes. The user interface was lousy. Scientists quickly created a custom display technology—the sweeping line and moving dots now called a PPI display. Pilots started using radar.

In some cases, as with the radar-controlled fuse on artillery shells mentioned earlier, Bush acted alone when he sensed a weak link. The Army initially paid little attention to the fuse, so Bush got on a plane and flew straight to battlefield headquarters in Europe. He was received by General Walter Bedell Smith, Eisenhower's chief of staff.

"What the devil are you doing over here?" Smith asked Bush. "Don't we have enough civilians in the theater without your joining?"

"I [came] over to a dense bed of ignorance," Bush replied, "to try to prevent the destruction of one of the best weapons of the war."

After this exchange, Bush reported, they got along swimmingly.

In other cases, Bush worked closely with FDR's secretary of war, Henry Stimson. When generals initially refused even to look at radar, Bush called Stimson. Stimson flew on an experimental plane equipped with the technology and watched as radar quickly sighted distant targets. The next day, the chiefs of both the Army and Air Force found identical notes on their desks:

> I've seen the new radar equipment. Why haven't you?

Key to that dynamic equilibrium—and Bush's ability to speak freely to generals—was support from the top. In the middle of managing a difficult conflict, Bush wrote, "I told FDR that he had handed me a hot potato, and I might have to bump some heads together. I remember well his answer. He said, 'You go ahead and bump, and I will back you up.'"

Not long after, one of the bumped heads came to FDR and launched into a tirade about Bush and his operation. The president, according to an aide who was present, was in the middle of signing letters. FDR paused for a while to listen, went back to signing letters, then said, "Look, Mac, I put that in Bush's hands. He's running it, and you get the hell out of here."

* * *

It may be helpful to visualize these first two rules, and what follows in the next several chapters, as shown below:

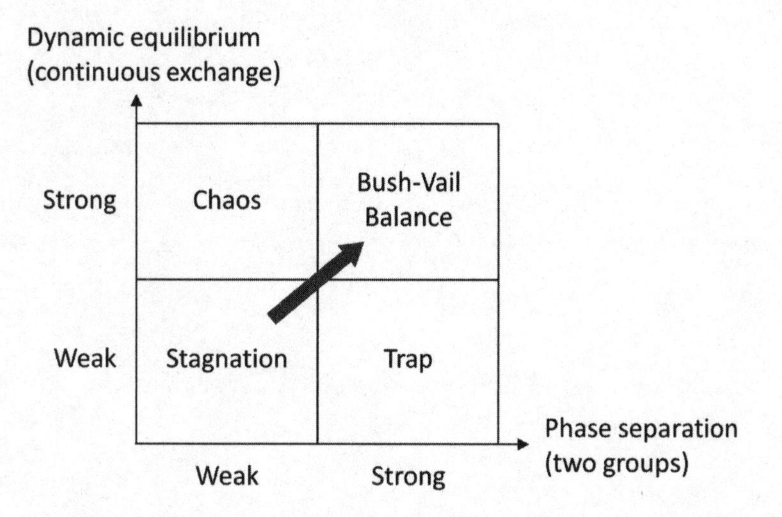

Bush and Vail succeeded in bringing stagnating organizations straight to the top-right quadrant: well-separated and equally strong loonshot and franchise groups (phase separation) continuously exchanging projects and ideas in both directions (dynamic equilibrium).

Many companies, however, especially when faced with a crisis, try to legislate creativity and innovation everywhere ("The CEO must be the CIO—the Chief Innovation Officer!"). This usually results in chaos, the top-left quadrant.

Not every phone operator has to be a champion innovator. Sometimes you just need them to answer the phone.

The most common trap, however, is to head straight to the bottom-right quadrant. As mentioned earlier, leaders proudly draw a box on an org chart, rent a new building, and hang a shingle advertising a new research lab. In chapters 3–5, we will see why that fails so frequently and how to get back on track to the top-right quadrant.

But first, we need to understand a little bit more about the *nature* of loonshots. Why do they need to be sheltered so carefully? Why are they so fragile?

2

The Surprising Fragility of the Loonshot

Akira Endo and the heart of stone

Sir James Black won the 1988 Nobel Prize in medicine for pioneering the modern approach to drug discovery. For five or six years, he periodically flew over from Britain and met with a small team of scientists from our biotech company to advise us on our research. Late one night, over some whiskey, after a marathon all-day science session, when I was close to collapsing from exhaustion and wondering how an 82-year-old man who had just traveled three thousand miles and talked all day could outlast me, I mumbled something about how hopeless some project seemed after a couple of failures in the lab.

Sir James leaned over, patted my knee, and said, "Ah, my boy—it's not a good drug unless it's been killed at least three times."

THE THREE DEATHS

Textbooks and colorful corporate brochures usually tell a different story, a happy history, a straight line from idea to cure. For example, most cancer research today centers on targeted therapies—drugs that hone in on cancer cells and leave normal cells alone. Books and magazine stories have glorified the rapid development of the first such "magic bullet" drug: Gleevec, one of the greatest breakthroughs in the history of treating cancer. Gleevec did in fact complete its clinical program astonishingly fast, still a record: thirty-five months from the first patient treated (June 1998) to FDA approval (May 2001). But before those trials, the scientist behind the drug, Brian Druker, was denied tenure because university research committees felt his work lacked potential; the major scientific journals rejected the paper describing his results; and Druker fought for years to convince the company that he eventually worked with to move the project forward. One executive at that company announced that Druker's project would advance "over my dead body."

In the real world, ideas are ridiculed, experiments fail, budgets are cut, and good people are fired for stupid reasons. Companies fall apart and their best projects remain buried, sometimes forever. The Three Deaths tells the honest history, as opposed to the revisionist history, of nearly every important breakthrough I'm aware of or have personally experienced (the Three often stretches to Four, Five, or Ten). The need to nurture and protect fragile loonshots so they can survive those stumbles and setbacks, whether self-inflicted or caused by others, is the central idea behind the systems of Bush and Vail.

As we will see, failing to understand the surprising fragility of the loonshot—assuming that the best ideas will blast through barriers, fueled by the power of their brilliance—can be a very expensive mistake. It can mean missing one of the most important discoveries in medicine of the century. And a $300 billion opportunity.

* * *

Tehran, November 28, 1943. Roosevelt, Churchill, and Stalin met together for the first time to discuss grand strategy: whether the Allies should launch a land invasion of Western Europe. Dinner was steak and potatoes. Stalin doodled wolf heads on a pad with a red pencil. Churchill lit cigars. At 10:30 p.m., midsentence, FDR "turned green and great drops of sweat began to bead off his face; he put a shaky hand to his forehead." He was wheeled to his room and placed in the care of his doctor, who attributed the attack to indigestion. Over the coming year, FDR's health would deteriorate rapidly. Friends noted his ravaged appearance and dramatic weight loss. On April 12, 1945, in what his physician described publicly as "a bolt out of the blue," FDR died of a sudden cerebral hemorrhage. To his medical advisors, however, it was anything but a bolt out of the blue. For years FDR had suffered from severe chronic heart disease.

At the time, heart disease was believed to be an inevitable result of aging, with no known cause or treatment. In 1768, addressing the Royal College of Physicians in London, William Heberden described a "disease which has hitherto hardly had a place or a name in medical books." He called it angina pectoris. "The termination," he said, "is remarkable. . . . The patients all suddenly fall down, and perish almost immediately." Heart attacks had been documented for thousands of years—"his heart died within him, and he became as a stone" (1 Samuel 25:37, ESV)—but Heberden's study of nearly a hundred patients was the first systematic attempt to understand and treat the underlying disease. Heberden concluded that he had little he could recommend to patients: quiet, liquor, and opium.

The death of the world's most famous American from heart disease galvanized support for research. In 1948, President Truman signed a bill creating the National Heart Institute. Modeled on Vannevar Bush's ideas in *Endless Frontier*, the center would award grants to scientists at universities, research labs, and hospitals to investigate the disease and possible treatments. Also included in the bill was funding for what eventually became the largest population study ever conducted: the Framingham Heart Study. The study's results, published in 1961 as "Factors of Risk in the Development of Coronary Heart Disease," established that elevated blood cholesterol conferred a high risk of heart attack or stroke (the paper was the origin of the term "risk factor").

The mortality rate from heart disease in the US—which had been growing since the beginning of the century—peaked in the late 1960s. Since then, the rate has decreased by roughly 75 percent, corresponding to well over 10 million lives saved over the past 50 years. Lifestyle changes—diet, exercise, the decline in smoking—are responsible for some of those gains. Much of the remainder is due to a drug isolated from a blue-green mold found in a grain store in Tokyo by a Japanese mushroom aficionado and microbiologist.

This is the story of that drug.

FUNGI DON'T RUN

The Framingham study ignited interest in cholesterol. Researchers launched dozens of clinical studies to evaluate whether new drugs or changes in diet could lower cholesterol and reduce heart attacks and strokes. In 1964,

Konrad Bloch and Feodor Lynen received the Nobel Prize for explaining how cholesterol is created and processed inside cells. And in 1966, a 33-year-old farmer's son, raised in a small mountain town in northern Japan, arrived in the United States determined to learn more about this new science. Akira Endo, a scientist from the food-processing division at the Japanese conglomerate Sankyo, joined a lab at New York's Albert Einstein College of Medicine that specialized in cholesterol research.

Endo arrived in the US just as the idea that diet could affect heart disease was taking off. A *Time* cover story described results from a new study by "the man most firmly at grips with the problem" of diet and health, Ancel Keys, a scientist at the University of Minnesota. His famous study of 10,000 people across seven countries confirmed that elevated blood cholesterol correlated with heart disease. But Keys went further, and implicated diet. Consuming fat, specifically saturated fat, he said, was the problem. Keys was not one for nuance. Obesity was "disgusting," he said. "Maybe if the idea got around again that obesity is immoral, the fat man would start to think." Keys's advocacy eventually led to official guidelines recommending low-fat, high-carbohydrate diets, despite the lack of any more rigorous evidence (it would be six decades before official guidelines backed away from that diet, now recognized not to be a good idea).

In a separate, less-publicized study, Keys compared heart disease rates between Japanese men living in Japan and those who had moved to Hawaii. The Hawaiian Japanese, who ate a Western diet, had much higher levels of cholesterol and heart disease than those who had stayed in Japan. In New York, Endo made the connection firsthand: He was surprised both by the high incidence of heart disease, and the rich American diet ("I saw many overweight people, like sumo wrestlers"). He concluded, like Keys, that as Japan became more Westernized, heart disease would become more common. He returned to Japan determined to find a drug that lowered cholesterol.

To find that drug, Endo turned to fungi—molds and mushrooms. As a child walking through the woods with his grandfather, Endo had noticed that a certain species of mushroom, safe for humans, was toxic to flies. Flies were everywhere after the war, so Endo, for a high school science project, showed that a broth made from the mushrooms was also toxic—proving that there was a water-soluble compound inside the mushrooms that killed flies.

Endo understood that fungi can't run, but they are great chemists. Mushrooms can't hop away from predators, so they secrete chemicals to discourage them (which is why so many mushrooms are toxic). Molds can't chase after food, so they secrete chemicals to make their hosts juicier and more nutritious. A juice-enhancing mold, in fact, had earned Endo his trip to New York. Endo had discovered that *Coniella diplodiella*, which causes a white rot on grapes, produces an enzyme that breaks down unwanted contaminants in juice and wine. The purifying enzyme became a hit for Sankyo, which rewarded Endo with the trip.

Since fungi are such great chemists, Endo reasoned, that's where he would begin his search. Bacteria, Endo knew, are natural predators of molds and mushrooms. To defend themselves, fungi have evolved many ways to kill bacteria. *Penicillium notatum*, for example, kills bacteria by secreting a compound that causes bacterial cell walls to collapse. That's how its extract, penicillin, works.

In New York, Endo had learned that many bacteria require cholesterol to survive. Could fungi secrete a chemical that kills their predators by blocking their cholesterol? In other words, Endo didn't want just any mold that kills mushrooms. He wanted a killer that used a specific weapon: a knife that surgically blocks cholesterol production. In the same way that forensic scientists use special tools to diagnose the weapon used by a killer, Endo would also need to use a special tool—similar to what forensic scientists do, but at a scale one million times smaller. It took Endo two years to build and perfect a state-of-the-art microscopic detection system.

Endo finally began screening fungi in April 1971. He tested just over six thousand species. In the summer of 1972, one sample lit up his system—what drug developers call a "hit." A blue-green mold, discovered growing on rice in a grain store in Kyoto, blocked a key enzyme needed to make cholesterol. The mold was *Penicillium citrinum*, the same genus that produced penicillin, but a different species. Within a year, Endo extracted the molecule that lowered cholesterol. He called it ML-236B. The drug is now known as mevastatin. It is the seed—the original—from which sprang Lipitor, Zocor, Crestor, and all the other statins. The statins would grow into the most widely prescribed drug franchise in history, saving millions of lives.

But first, Endo's drug had to survive the Three Deaths.

SAVED BY CHICKENS

Shortly after Endo began screening fungi in Japan, trials in the US studying the effects of lowering cholesterol, which had been launched with great enthusiasm several years earlier, landed with a thud. Dietary intervention showed little, if any, benefit. One widely publicized editorial in the *New England Journal of Medicine*, titled "Diet-Heart: End of an Era," buried the idea of any connection, describing efforts to lower cholesterol as "a fund raiser for the Heart Association, and busy work for thousands of fat chemists."

Trials evaluating drugs that lowered cholesterol did even worse than the diet trials. Three of the most widely studied drugs were found to *increase* the overall death rate in trials. Another clearly caused cataracts. An editorial in the *British Medical Journal,* written by one of England's most respected cardiologists, summed up the prevailing view: "All well-controlled trials of cholesterol-reducing diets and drugs have failed to reduce coronary (CHD) mortality and morbidity." Another editorial declared, "The evidence that eliminating risk factors will eliminate heart disease adds up to little more than zero."

Since normal cell function requires cholesterol, writers of prestigious scientific reviews invoked commonsense biology to explain the failures. *Any* drug that lowered cholesterol must be dangerous because it would interrupt normal cell function. Academics lost interest and most companies gave up. Endo presented his promising results for mevastatin at a conference around that time. But by then, the idea of lowering cholesterol had been crushed by consensus. Almost no one came to hear his talk. He left the meeting dejected. (Death #1.)

Endo's small team at Sankyo faced intense skepticism from management and colleagues. Expecting the worst, Endo discussed with his wife whether she would be willing to support their family on her income alone if he were fired. She agreed. He drafted a letter of resignation and carried it in his pocket, ready to produce it at any time if asked. He would leave with dignity.

To Endo's surprise, no one asked for the letter. The goodwill he had built up from his prior successes, along with a tolerant supervisor, shielded him, at least for the moment. Mevastatin soon reached a crucial stage: testing in live animals. The honor of being first is generally awarded to rodents. With great excitement, the team gave the drug to rats and saw . . . nothing. No cholesterol lowering. In the world of drug discovery, failure in standard animal studies nearly always kills a project. With those results, Endo recalled years later, there was no hope of convincing biologists at Sankyo to continue evaluating the drug. (Death #2.)

Endo pleaded for and was granted more time to figure out *why* his drug didn't work. At a bar near his lab, he ran into Noritoshi Kitano, a colleague from a different department who worked with chickens. After a few drinks, Kitano confided that his chickens would make a nice dish of yakitori when his research project ended the following month. It occurred to Endo that hens might have high blood cholesterol, since their eggs have so much of it. Higher starting levels of cholesterol could make the effects of his drug easier to detect. So Endo convinced Kitano to curb his appetite, temporarily, and test mevastatin on some spare hens. They began without formal approval. When I asked Endo whether they were able to keep the experiment secret, he laughed and said, "Hens cluck. It is impossible to hide them."

The results were spectacular. Mevastatin decreased cholesterol by nearly half, triglycerides by even more, with no ill effects. Much later, scientists learned that rats have mostly HDL ("good cholesterol") circulating in their blood, and very little LDL, the "bad cholesterol" that contributes to heart disease. Which means that rats are a poor choice for evaluating statins, which lower just the LDL. Chickens have both types, like humans.

Around the time that Endo discovered his drug worked in chickens—and, soon afterward, in dogs and monkeys—two physician-scientists at the University of Texas in Dallas had just begun what would soon become an extraordinary scientific duet. Michael Brown and Joseph Goldstein met as young residents at the Massachusetts General Hospital in Boston in 1966. Both continued their training at the National Institutes of Health in Maryland in 1968. At the NIH, Goldstein was asked to care for a six-year-old boy and his eight-year-old sister suffering from repeated heart attacks. The siblings were diagnosed with familial hypercholesterolemia (FH), an inherited genetic disease.

Roughly one in five hundred people are born with a defective gene for a protein that siphons LDL cholesterol out of the blood. The weakened ability to pump cholesterol out of the blood causes cholesterol levels to rise to twice normal levels. Patients typically begin having heart attacks in their thirties. One in a million people inherit the defective gene from both parents and are born with FH, like the children Goldstein saw. They have up to ten times the normal levels of cholesterol in the blood and often begin having heart attacks early in childhood. Brown and Goldstein decided to work together to find a treatment. They ended up at the University of Texas, published their first paper together in 1973, and have published over *five hundred* papers together in the past forty years (alternating the order of their names on each, from "Brown-Goldstein" to "Goldstein-Brown"). They have been called the Gilbert and Sullivan of medicine.

After Brown and Goldstein arrived in Texas, they subscribed to a computer-based service that alerted them to published articles citing their work (not uncommon in the pre-internet era). In July 1976, the service notified them that one Akira Endo in Tokyo had published an article in a Japanese scientific journal reviewing the results from one of their papers. They couldn't read the Japanese words, but they recognized the figures from their paper. They were delighted that their work had crossed overseas and added Endo to their author screen. Several months later, the service alerted them to two new articles by Endo, published in December 1976. The papers described his discovery of mevastatin. Brown and Goldstein immediately grasped the importance for patients with FH.

Goldstein wrote to Endo and asked for a sample of his drug, which he promptly provided. The Texas scientists confirmed Endo's laboratory results in their own lab and encouraged him to test the drug in patients. That same summer, in 1977, a physician in Japan, Akira Yamamoto, also read about Endo's work. Yamamoto called Endo about an 18-year-old girl with severe FH. Her condition was serious. Endo, encouraged by the support of Brown and Goldstein, agreed that his drug was worth trying. On February 2, 1978, Yamamoto's patient, known later in the literature only by her initials, S.S., became the first person treated with a statin.

Two weeks into the trial, Yamamoto called Endo at home at midnight: S.S.'s cholesterol had dropped by 30 percent—the drug worked! The trial succeeded and the drug became the first big hope for patients with dangerously high cholesterol. Sankyo launched an official clinical program, which expanded into a large, twelve-hospital study in 1979. The mevastatin results attracted interest around the world. In May 1980, a special workshop on mevastatin was held in Italy. Eight Japanese physicians who were treating patients with mevastatin attended and presented.

Endo was satisfied that his drug project was secure in the hands of capable physicians who would move it forward through clinical trials and regulatory approval. Having had enough of company battles, he retired from Sankyo and accepted a university research and teaching position in Tokyo.

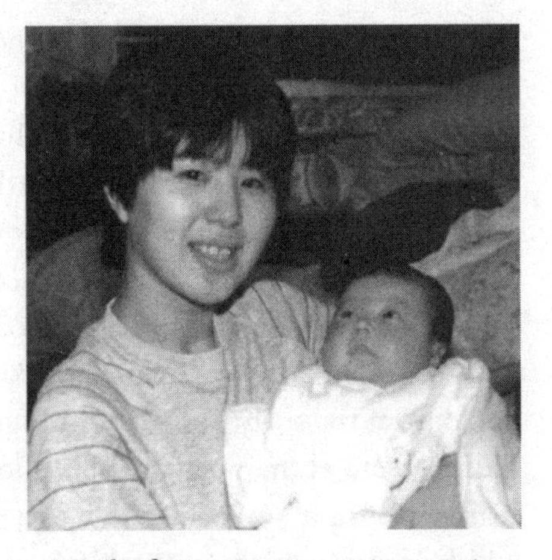

S.S., the first patient to receive a statin, seven years later, holding her baby

The worldwide enthusiasm for mevastatin, however, was short-lived. Three months after the meeting in Italy, results from a safety study conducted by Sankyo delivered a knockout punch: high doses of mevastatin appeared to cause cancer in dogs. Sankyo had had enough. The company stopped the trial and mevastatin research. Rumors about the cancer side effect spread quickly. Other companies and research centers terminated their statin research. Although Endo suspected the dog study was flawed, he could only watch from a distance as his program collapsed. (Death #3.)

A $90 BILLION "COINCIDENCE"

That might have been the end of the statins, except for a surprising discovery from a parallel research program. Two years earlier, the pharma giant Merck had *also* initiated a screen of fungi, had *also* discovered an inhibitor of the same enzyme that Endo discovered, and had *also* found that it worked well in lowering cholesterol. Remarkably, the Merck compound differed from Endo's by just four atoms.

Also remarkable was that Merck scientists discovered their drug, in November 1978, within *days* of starting their program, compared to the years that Endo had invested. Roy Vagelos, the head of Merck Research Lab at the time, described the "sudden" discovery as "unbelievable." In his memoir, the highlight of which is his team's discovery of the statins, Vagelos describes the energy of competing with Sankyo: "As the pace picked up, the excitement steadily mounted. The competition with Sankyo heightened the thrill of discovery."

A competition, however, generally requires participants who believe they are competing. Every account that I have read by Merck scientists about the discovery of the statins has omitted a relevant detail: two and a half years *before* Merck's sudden discovery, Merck approached Endo and his team to *collaborate*, rather than compete, requesting access to their most confidential proprietary data. Based on Merck's assurances in letters dated from the spring of 1966 through the fall of 1968 ("It seems evident that a practical therapeutic application will develop from Dr. Endo's research program"; "We hope that as a result of these exchanges, a product will be found which is suitable for license"), Endo and his team, with Sankyo's approval, not only provided samples of Endo's drug for testing but also shared results from crucial experiments, including the drug's biochemistry, pharmacology, efficacy, and toxicity—priceless information. The letters document Endo and his team visiting Merck's labs in New Jersey, receiving the Merck scientists in Japan, and responding to detailed questions from Merck scientists about the drug. In that context, Merck's "sudden" discovery two years later of a nearly identical drug becomes a bit less . . . "unbelievable."

Around the time that Sankyo terminated its program, Vagelos heard the rumors that Sankyo's drug caused cancer in dogs. So Vagelos, aware that their two compounds were very similar, also terminated Merck's program. Those rumored results, however, were never published or confirmed, either at the time or in many subsequent studies. Endo, by then settled in at Tokyo Noko University, was skeptical and asked Sankyo to share the data with him. The company declined. Brown and Goldstein, at the University of Texas, were also skeptical. They soon proved that the ultra-high doses used in dogs could lead to a harmless condition that looked like cancer but wasn't: a false positive. Together with several other physicians, and with support from the FDA, Brown and Goldstein pressured Merck to restart its program.

Merck accepted their arguments and began new safety studies. After those studies showed no hints of the drug causing cancer in dogs, Merck then began the large clinical trials required by the FDA to establish the safety and efficacy of a drug. The results were strikingly positive, consistent with the earliest data observed by Endo and Yamamoto in their clin-

Akira Endo

ical studies. In February 1987, an FDA advisory panel unanimously recommended approving the first statin: Merck's Mevacor.

The early studies by Merck and various physician groups showed only that statins could lower dangerously high cholesterol levels. That was an important and encouraging *marker* of benefit, but not yet definitive evidence of improvement in health. Hundreds of researchers subsequently launched dozens of randomized, controlled clinical trials. Those trials, which to date have enrolled over 100,000 participants, have established statins as one of the great medical breakthroughs of the twentieth century. Statins reduce heart attacks and strokes and extend survival, not only in patients who have survived heart attacks (secondary prevention) but also in at-risk patients who have never yet experienced heart attacks (primary prevention). In the US, statins prevent approximately half a million heart attacks and strokes *each year.* One recent editorial in the *New England Journal of Medicine* concluded, "Few drugs have had such a dramatic effect on health outcomes."

Mevacor and its successor Zocor became the most successful drugs in Merck's history. Cumulative sales of Merck's statin franchise have exceeded $90 billion. Cumulative sales from all statins have exceeded $300 billion.

Vagelos was promoted from head of research to CEO of Merck in 1985. From 1987 through 1993, the company won the top honor in the *Forbes* Most Admired Company competition. In 1985, for their work on cholesterol, Brown and Goldstein were jointly awarded the Nobel Prize.

Endo's contributions, on the other hand, have gone largely unrecognized outside the small field of cardiology specialists. Within that world, however, Endo has received some belated appreciation. In 2008, he was awarded the prestigious Lasker-DeBakey prize for discovering the statins. Brown and Goldstein dedicated a recent historical review "to Akira Endo, discoverer of a 'penicillin' for cholesterol," and concluded, "The millions of people whose lives will be extended through statin therapy owe it all to Akira Endo and his search through fungal extracts at the Sankyo Co."

* * *

Endo's story is more than a wild anecdote. The twisted paths leading to great discoveries are the rule rather than the exception. And so are their revisionist histories: victors don't just write history; they rewrite history.

Endo's journey, from start to ultimate validation with the FDA approval of the first statin, lasted sixteen years. Sir James Black's journey through the Three Deaths (inventing beta-blockers, a class of drugs for treating high blood pressure) lasted seven years.

One man survived a far longer and more personally trying journey, championing a loonshot that was ridiculed by colleagues around the world, including many from his own institution, for *thirty-two years*.

He is the person from whom I learned the most about the long journeys of great discoveries. In closing this chapter, I'll briefly describe his story, and then explain what his story and Endo's together tell us about surviving the Three Deaths.

COUNTING THE ARROWS IN YOUR ASS

Sometime in 2001 or 2002 I asked a friend, a biologist at Harvard, about the man behind a radical idea for treating cancer. My company and I were considering working with him. My friend, well known in the research world for his good nature and generosity, looked embarrassed and mumbled that we might want to stay away because "people can't reproduce his data," and then quickly changed the subject. That was the word at the time on Judah Folkman.

In 1971, Folkman had proposed that cancer cells interact with their hosts, sending out signals that trick surrounding tissues into preparing the local environment for a tumor to grow. Tumors, for example, need blood vessels to bring in oxygen and other nutrients, just like a home needs pipes to bring in water and gas. Folkman suggested that cancer cells signal surrounding tissues to produce those blood vessels. His idea was to design a new kind of drug, one that blocks those signals and destroys those pipes. In other words, a drug that starves a tumor.

At the time, the only approach to treating cancer was chemotherapy: flood tumors with poison, the more the better, as long as you don't kill the patient at the same time. The idea of interrupting some mysterious communication channel between tumors and their surrounding tissue was greeted with sneers. It didn't help that Folkman was a pediatric surgeon and came from outside the cozy club of research PhDs. At scientific meetings he would get up to speak, he said, and the room would empty out: "Everybody had to go to the bathroom at once." One year, the criticism was so intense that Boston Children's Hospital, where he worked, convened an external committee to review his science. The committee judged his work to be of little to no value. He was asked to resign as chief of surgery if he chose to continue that research. In a speech years later, Folkman said, "If you think you have suffered ridicule, write to me and I will send you pink sheets from referees [reviewers of his rejected articles] and grant committees from the mid-seventies that include the word 'clown.'"

For three decades, there was a roughly seven-year cycle between deaths and spectacular rebirths of Folkman's idea. In 1998, for example, a promising drug from Folkman's lab was shown to eradicate tumors in mice. A page one *New York Times* story quoted the Nobel laureate James Watson saying, "Judah will cure cancer in two years" (Watson later challenged the quote). Media coverage exploded. Reporters compared Folkman to Alexander Fleming and Louis Pasteur; a Pulitzer Prize–winning columnist who had been diagnosed with colon cancer wrote a column announcing, "Maybe we don't have to die"; and patients besieged Folkman's hospital for access to the drug, which was not yet in clinical trials. As with most new ideas in drug discovery, the first drug didn't pan out. Interest plummeted.

After a few such cycles, most of the scientific community wrote off both Folkman and his ideas. He would hear people laughing in the corner during his presentations. Colleagues would say, "Oh, I see Folkman has cured cancer again." Occasionally, scientists would stand up at the end of his talks and announce that his idea could never work. Folkman's response: "I have a little book that I carry . . . Will you sign for me? Because you're so sure, I can just publish your remarks directly and save a lot of government and taxpayers' money, and we won't do the experiments . . . We'll just say it won't work." But still he would go home depressed.

At one point, Folkman discussed with his wife, Paula, whether he should quit research, close the lab, and return full-time to surgery. With Paula's encouragement, which he later called "Spouse Activation Factor" (SAF), Folkman did the opposite. He quit clinical work and began full-time research. He recruited a handful of star students, overcoming warnings they had heard to stay away from Folkman and his work "by reminding them that they were so good that even if things didn't work out and they left after a year, their careers wouldn't be harmed," he recalled years later. Folkman joined them in the lab, working nights and weekends.

On June 1, 2003, in front of a packed auditorium in Chicago's McCormick Place convention center, 32 years after Folkman first proposed a new type of cancer therapy, well after Folkman's arguments and pleas had faded from many memories, Dr. Herbert Hurwitz, an oncologist from Duke University, unveiled new results from a drug called Avastin, designed based on Folkman's ideas. In a clinical trial that enrolled 813 patients, Avastin demonstrated the best results ever seen for prolonging survival in patients with colon cancer. When Hurwitz showed the survival data, the room burst into applause. It was instantly clear that the drug and Folkman's ideas would transform the treatment of cancer.

One audience member said, "I only wish Dr. Folkman were alive to see this." Sitting nearby, Folkman just smiled.

The drug was rapidly approved by the FDA; dozens of companies and hundreds of research labs jumped into the field; and today the idea of interrupting the dialogue between the tumor and its host environment underlies targeted therapy, immunotherapy, and nearly every active cancer research program. The company that developed Avastin was called Genentech. Between the day the company first announced the data and the day the FDA approved the drug, its market value increased by $38 billion, a rough measure of the value of the drug. (Folkman owned no stock in the company; he routinely donated any financial stakes and prize money he received to his hospital.)

Later, Folkman would say, "You can tell a leader by counting the number of arrows in his ass."

I ended up overlooking my friend's mumbled suggestion and worked happily with Judah for the last seven or so years of his life. I miss him.

§

LESSONS FROM THE SURPRISING FRAGILITY

Beware the False Fail

The Endo and Folkman stories illustrate not only the Three Deaths but also a specific *type* of death, one common to loonshots. The failure of Endo's drug in rat models (Death #2), for example, nearly terminated his program at Sankyo. The same failure *permanently* killed a similar program at another company, Beecham Pharmaceuticals. Beecham later merged with SmithKline & French, and then Glaxo Wellcome, becoming today's GlaxoSmithKline. Had Beecham persisted, they might have shared in the $300 billion of revenues from statins. Even a small piece of $300 billion is pretty good. But they gave up and ended up with nothing.

The negative result in the rat experiment was a False Fail—a result mistakenly attributed to the loonshot but actually a flaw in the test. Sankyo persisted through that fail, because of Endo. It was winning the race. Because of Endo, it was the first to discover a statin, the first to patent a statin, the first to test statins in humans, and the first to see clinical benefit in patients. But it gave up at the next False Fail, after Endo had left: the spurious results in dogs. The company handed its share of $300 billion to Merck.

We will see the False Fail over and over, both in science and in business. There are many reasons projects can die: funding dwindles, a competitor wins, the market changes, a key person leaves. But the False Fail is common to loonshots. Risks of this type of death can never be fully eliminated—negative results don't come with a neon sign that lights up "your idea is flawed" or "your test is flawed." But those risks can be reduced, which is exactly what Endo and Folkman did, as we will discuss below. People may think of Endo and Folkman as great inventors, but arguably their greatest skill was investigating failure. They learned to separate False Fails from true fails.

Skill in investigating failure not only separates good scientists from great scientists but also good businessmen from great businessmen. In 2004, for example, when Facebook launched, many social networks had tried and failed to win the loyalty of users who hopped from one network to the next: Classmates, Sixdegrees, Care2, AsianAvenue, BlackPlanet, KiwiBox, LiveJournal, StumbleUpon, Elfwood, Meetup, Dodgeball, Delicious, Tribe, Hub Culture, hi5, aSmallWorld, and others. As Mark Zuckerberg met with investors to raise funds for his new startup, users were just beginning to abandon the most recent social network success story, Friendster, for MySpace. Most investors concluded the websites were like clothing fads. Users switched networks like they switched jeans. Investors passed.

Peter Thiel and Ken Howery at Founders Fund, however, reached out to their friends behind the scenes at Friendster. They dug into *why* users were leaving the site. Like other users, Thiel and Howery knew that Friendster crashed often. They also knew that the team behind Friendster had received, and ignored, crucial advice on how to scale their site—how to transform a system built for a few thousand users into one that could support millions of users. They asked for and received a copy of Friendster's data on user retention. They were stunned by how long users stayed with the site, *despite* the irritating crashes.

They concluded that users weren't leaving because social networks were weak business models, like clothing brands. They were leaving because of a software glitch. It was a False Fail.

Thiel wrote Zuckerberg a check for $500,000. Eight years later, he sold most of his stake in Facebook for roughly a billion dollars.

Thiel saw past the False Fail of Friendster, just as Endo saw past the False Fails of the statins and Folkman saw past the False Fails of his blood-vessel inhibitors.

Create project champions

Fragile projects need strong hands. After Endo left Sankyo, for example, the company's statin program withered and eventually collapsed. There was no one internally to investigate and answer False Fails, no one to protect the program from critics with other agendas who wanted its budget for their own programs.

Endo was both the inventor of an idea and its skilled champion, as was Judah Folkman. But that combination is rare. It's natural to assume that the inventor of an idea should also be its chief promoter and defender. But the best inventors do not necessarily make the best champions. The roles require different skills not often found in the same person.

In chapter 1, we saw that Hoyt Taylor and his team, who discovered the principles of radar in the 1920s, were good inventors. But they were lousy champions. They didn't know how to package and promote a new idea, how to convince skeptical leaders, how to build support inside a reluctant organization.

Accounts of the origin of radar in the US military nearly always omit the person who deserves most of the credit for the US catching up in time to make a difference: Lieutenant (later Admiral) William "Deak" Parsons. A career Navy officer who read *Reviews of Modern Physics* in his spare time, Parsons returned from his second tour at sea in the spring of 1933 at age 31. The Bureau of Ordnance assigned him as its liaison to the "small, little-known bit" of the Navy known as the Naval Research Laboratory. What he found blew him away:

> Parsons immediately grasped the military possibilities of the experimental work described by Taylor. . . .
>
> A radio echo device that could detect aircraft beyond the limits of human sight could protect ships and harbors from surprise attacks, save lives, perhaps turn the tides of battles. Yet to Parsons' dismay he learned that the exploratory research on this radical concept was limping along without any priority, a part-time effort of only two professionals. Neither the Navy nor the scientists seemed to share Parsons' excitement. . . .
>
> No one seemed to recognize what was immediately obvious to him: that the radio-echo discovery could revolutionize naval weaponry.

After pumping Taylor and his crew for more information, Parsons immediately circulated a proposal for $5,000 in funding. He was stunned when it was rejected. The same skepticism that silenced Taylor fired up Parsons. "With the persistence of a door-to-door salesman," Parsons took the idea to every head of desk in the Navy and made the case, putting his career at risk. Parsons reenergized the scientists at the naval lab—inspiring Taylor to assign the first engineer dedicated to the project (Robert Page, who made the critical breakthrough of using a pulsed rather than continuous signal)—and he convinced top military brass to stand up and fight for the project. He prodded and poked until the sleeping bear woke.

Years later, both Vannevar Bush and Rear Admiral Frederick Entwistle, who oversaw antiaircraft protection for ships during the war, credited Parsons for operational radar being ready by the start of World War II.

That's a project champion.

Many of the best biotech and pharma companies today have learned to separate the roles of inventor and champion. They train people for the project champion job—the Deak Parsons skill-set—and elevate their authority. It goes against the grain. On the creative side, inventors (artists) often believe that their work should speak for itself. Most find any kind of promotion distasteful. On the business side, line managers (soldiers) don't see the need for someone who doesn't make or sell stuff—for someone whose job is simply to promote an idea internally. But great project champions are much more than promoters. They are bilingual specialists, fluent in both artist-speak and soldier-speak, who can bring the two sides together.

Although creating the role may induce some eye-rolling, the teams or companies that do it well will reduce the risk of what happened with radar at the Navy. They will avoid burying a great idea for lack of a great champion.

LSC: Listen to the Suck with Curiosity

On every setback or rejection I experience, which occurs often, I try to remind myself of a third lesson from the fragility of loonshots. It's how Endo, Folkman, and Thiel got past False Fails. I think of it as Listening to the Suck with Curiosity (LSC)—overcoming the urge to defend and dismiss when attacked and instead investigating failure with an open mind.*

While others gave up when, for example, the animal models didn't work, Endo asked *why* they didn't work, and set about testing ideas. Well before he convinced his friend with the pre-yakitori chickens to administer last-rite statin, Endo had spent many months on experiments trying

* The word choice is borrowed from a phrase made famous by Sheryl Sandberg.

to understand *why* his drug didn't behave as he expected. He already suspected what's called a species difference (when a drug behaves very differently across animal species). He knew to act quickly when the opportunity appeared.

Where others assumed Friendster was yet another example of a social network fad, Thiel and Howery investigated more deeply *why* users were leaving and found a contrarian answer, in which they had confidence. Contrarian answers, with confidence, create very attractive investments.

I mentioned earlier that a biologist friend had advised me to stay away from Judah Folkman because "people couldn't reproduce his data." In fact, initially, some people really couldn't reproduce Folkman's data. Shortly after Folkman published a landmark paper in 1997, other labs wrote to ask for materials and instructions to confirm and expand on his results. He promptly sent both. At some labs, the experiments didn't work (at others, they succeeded). A reporter heard about the failures. The headline of his 1998 *Wall Street Journal* article read, "Novel Cancer Approach Stumbles as Others Fail to Repeat Successes." In academia, irreproducible results end careers—especially when they make headlines in national newspapers.

Rather than lash out at critics, however, Folkman investigated. He tried to understand exactly what other labs were doing, *why* their experiments were failing. Eventually, he discovered that some of the sensitive sample materials his lab sent out were being damaged by the freezing process used to ship them long distances. He changed how samples were shipped. Experiments started to work, and labs around the country came around in support of his work.

It was from Judah that I first saw LSC in action, consistently. He would (usually) overcome the impulse to challenge his challengers when attacked. He kept an open mind and quietly investigated, with genuine interest and a desire to learn.

Why the C in LSC? I have been a frequent, and occasionally willing, recipient of management training workshops and sensitivity sessions. The "active listening" mantra is drilled into your head. Repeat back what you've just heard to show you understand. But when investors reject your pitch, customers reject your product, or a partner walks away, indicating that you received that message is not, by itself, enough. If you've poured

Michael Brown (L), Akira Endo, and Joseph Goldstein (R)

your soul into a project, the temptation to dismiss bad outcomes is high. What you crave is reassurance that you're on the right track. So you ignore or attack your challengers and turn for reassurance to your friends, mentors, mother.

LSC means not only listening for the Suck and acknowledging receipt but also probing beneath the surface, with genuine curiosity, *why* something isn't working, *why* people are not buying. It's hard to hear that no one likes your baby. It's even harder to keep asking why.

LSC is also the answer I give to the most anguished question I often hear from entrepreneurs, or any champion of any kind of loonshot. The question usually comes up only late at night, after a few drinks, after a discussion of daily struggles winds down and the conversation turns existential, when the weariness from years of blows to the body seeps out. "How do I know when to give up?"

How *does* one tell the difference between persistence and stubbornness?

LSC, for me, is a signal. When someone challenges the project you've invested years in, do you defend with anger or investigate with genuine curiosity?

I find it's when I question the least that I need to worry the most.

3

The Two Types of Loonshots: Trippe vs. Crandall

Jet engines vs. frequent fliers

In 1968, when its founder and chief executive retired, Pan Am was the largest and most profitable airline in the world, and the most recognized brand after Coca-Cola. It had been the first American airline to fly transatlantic, the first to fly transpacific, the first to complete a round-the-world flight, the first to operate jets. James Bond flew Pan Am in *From Russia with Love.* The Beatles held their first American press conference in front of a Pan Am Clipper. Pan Am captains were asked for autographs, like movie stars. The Pan Am Building in New York, completed in 1963 with 25-foot blue Pan Am globes on top, was the largest office building in the world. The 1968 movie *2001: A Space Odyssey* showed an elegant Pan Am cabin attendant, gliding through a Pan Am spaceship, wearing Pan Am slippers, serving tasty treats. Back on earth, Pan Am had begun accepting reservations for the first commercial flights to the moon.

The next year, Pan Am posted its first loss. Over the next 22 years, Pan Am lost money in every year but four.

On the morning of December 4, 1991, Captain Mark Pyle of the Boeing 727 *Pan American Clipper Goodwill* was awaiting takeoff on the tarmac in Barbados when the airline's station manager walked toward his plane and signaled him to meet in the cockpit. Minutes later, Pyle emerged and told the flight attendants: Pan Am had ceased operations. They broke down in tears. A few hours later Pyle flew one low pass over Miami airport and then landed. As the plane slowly taxied toward the gate, ramp workers and airline staff stood at attention and saluted. Water cannons fired streams over the plane. A few months later, the blue globes came down from the building in New York, replaced by the white block letters of an insurance company.

What happened?

* * *

In the previous two chapters, we saw the *needs* behind the Bush-Vail system. We need to protect and nurture loonshots, because of their surprising fragility. We need to balance loonshots and franchises, because they strengthen each other. Those needs gave rise to the first two rules: phase separation and dynamic equilibrium.

In this chapter and the next two, we will see a third need: the need to distinguish between two *types* of loonshots.

Missing one kind of loonshot brought down the world's most exciting airline company. Missing the other kind brought down the world's most exciting consumer technology company. Both companies learned, irreversibly, what Vannevar Bush and Theodore Vail already knew.

Missing loonshots can be fatal.

THE TWO TYPES

Let's call a surprising breakthrough in *product*—a technology that was widely dismissed before ultimately triumphing—a *P-type* loonshot. "To the business world the telephone was just a toy," read a biography of Theodore Vail in 1921. Investors "smiled or made some facetious remark when invited to invest in the stock." That would be the Bell Telephone Company. It would grow into the most valuable company in the country, more dominant at its peak than Apple, Microsoft, and GE, at their respective peaks, combined.

With P-type loonshots, people say, "There's no way that could ever work" or "There's no way that will ever catch on." And then it does.

Let's call a surprising breakthrough in *strategy*—a new way of doing business, or a new application of an existing product, which involves no new technologies—an *S-type* loonshot. Sam Walton located his stores far from major cities, supersized them, and sold $1.20 women's underwear for $1.00. There were no new technologies. He found a different way of delivering the same products, slightly cheaper. Walmart, in 2018, is the largest retailer on the planet. If it were a country, it would rank #25 in the world by GDP. Its top former competitors—Woolworth, Federated, Montgomery Ward, Gibson's, Ames—are long gone.

With S-type loonshots, people say, "There's no way that could ever make money." And then it does.

Facebook didn't invent social networks and Google didn't invent search, just like Walmart didn't invent selling stuff cheaply. Early investors passed on Facebook because everyone knew there was no money to be made in social networks. They passed on Google because everyone knew there was no money to be made in search. Both succeeded because of small changes in strategy no one thought would amount to much. Both succeeded because of S-type loonshots.*

Deaths from P-type loonshots tend to be quick and dramatic. A flashy new technology appears (streaming video), it quickly displaces what came before (rentals), champions emerge (Netflix, Amazon), and the old guard crumbles (Blockbuster). Deaths from S-type loonshots tend to be more gradual and less obvious. It took three decades for Walmart to dominate retail and variety stores to fade away. And no one could quite figure out what Walmart was doing, or why it kept winning.

S-type loonshots are so difficult to spot and understand, even in hindsight, because they are so often masked by the complex behaviors of buyers, sellers, and markets. In science, complexities often mask deep truths: mountains of noise conceal a pebble of signal. We design experiments in labs to strip away those complexities and reveal the hidden truths. But occasionally, rare events in nature do the job for us.

A solar eclipse is an example of a rare natural experiment. During a solar eclipse, the moon blocks the light from the sun, allowing us to see the faint light from distant stars during the day. In 1919, a British team measured the bending of distant starlight by the sun during a solar eclipse. They showed that Einstein's theory of gravity, proposed only four years earlier, explained the deflection of light much better than Newton's theory.

---

* For business theorists: The two types of loonshots are unrelated to what Louis Galambos in 1992 called "adaptive" vs. "formative" innovations, and Clayton Christensen in 1997 called "sustaining" vs. "disruptive" innovations. For the distinction, see the afterword.

In 1978, when it deregulated the airline industry, Congress created the business world's equivalent of a solar eclipse: a rare-event experiment.

For fifty years, the federal government had regulated where airlines could fly and what they could charge, down to the tiniest details: the price of a cocktail, the rental cost of a movie headset. Suddenly removing these restrictions unleashed a tidal wave of S-type loonshots, small shifts in strategy. Those changes were not glamorous. They were kind of nerdy: a frequent flier program, a new system of flying through hubs rather than flying direct, a computerized reservation system for travel agents. P-type loonshots—jet engines, jumbo planes—make headlines. Small changes in strategy are barely noticed. Deregulation, for a brief moment, let the faint, hidden light from S-type loonshots shine through.

Most of those S-type loonshots were invented or perfected by American Airlines CEO Bob Crandall. Crandall was a master S-type innovator. Most of the industry's P-type loonshots were invented or perfected by Pan Am's founder and CEO, Juan Trippe. Trippe was a master P-type innovator. Between 1978 and 2008, deregulation helped drive 170 airlines out of business or into bankruptcy, including Pan Am and every major US carrier, except for one—American Airlines.

Few industries today are as regulated as airlines were before 1978. But sudden shocks happen all the time. When Google announced one morning that it would give away its new mobile phone operating system—Android—for free, the rules of the mobile world suddenly changed. That announcement, like deregulation, triggered a tidal wave of S-type loonshots that caught every unprepared company by surprise.

Which is why learning to nurture the more subtle S-type loonshots—not just the shiny P-type loonshots—matters. Most people, like most teams and companies, have a blind side. And the subtle is much easier to miss than the shiny.

If you are a creative or an entrepreneur, learning how to be good at both types of loonshots can help you expand your idea. It can help you transform something good into something great. Google, for example, began with a new algorithm for ranking internet search results, a nice P-type loonshot. But it was the eighteenth search engine. It added several clever S-type loonshots to attract advertisers. Those S-type loonshots helped it grow into the dominant website in the world.

If you are an industry challenger, learning how to be good at both can help you defeat bigger, stronger competitors, like a middleweight firing off a surprising left hook to knock out a heavyweight.

And if you are an astonishingly successful innovator, if you have built a wondrous empire, you need to learn how to watch your blind side—how to spot the loonshots blazing right toward you.

You need to know how not to be Pan Am.

JT AND CRANDO

Pan Am, for nearly its entire existence, was Juan Terry Trippe, just like American Airlines, for eighteen years, was Robert Lloyd Crandall.

Trippe, who founded Pan Am in 1929, hated his Spanish-sounding first name (honoring his mother's half sister, Juanita) and switched to JT. He was the son of a New York investment banker whose family traced its American roots to 1663; he grew up with Whitneys, Vanderbilts, and Rockefellers; he attended Yale, playing football and golf. He didn't speak a word of Spanish. But when Pan Am first took shape in Latin America, he switched back to Juan and hired a bilingual assistant who could write letters, in his name, in fluent Spanish, to local presidents and dictators. Within five years, he owned the air over the continent. Within ten years, he dominated international flight. Franklin D. Roosevelt described him as "the most fascinating Yale gangster I ever met." A colleague described him as "the politest and least compassionate man I have ever known."

Where Trippe was polite patrician cool, Crandall was raw chain-smoking testosterone.

Crandall believed in a philosophy he called competitive anger: "You ought to be angry at your opponent, and you ought to be angry with yourself if you don't win." He was called Attila the Hun, Bob the Butcher, Darth Vader, and—in case the message was still not clear—Fang (he has prominent canine teeth). On weekends, he'd go to work and leave notes on desks: "I was here. Where were you?" In a 1987 company video, he burst onto the screen in military uniform, face paint, and bandanna, carrying a plastic toy machine gun—Crando. One biographer noted his obsession with order: "If he noticed his wife's purse sitting on the kitchen counter, he might pry it open for inspection; finding it in the same condition as any purse, he would dump it out and reorganize it, throwing away the bits of grit that had accumulated in the creases at the bottom. 'It drives her batshit,' he would remark with a raspy nicotine cackle."

Although their styles differed, both Trippe and Crandall were ruthless and ambitious. Both wanted to dominate the world skies. Both hated deregulation, when it finally arrived in 1978. This is Bob Crandall at the US Senate hearings, explaining his views to the economists and lawyers in the room: "You f—king academic eggheads! You're going to wreck this industry!"

But after deregulation, as mentioned earlier, Crandall's airline survived and thrived, while Trippe's withered and died. It seems like the opposite should have happened. Unlike Trippe, Crandall was not an aviation guy. He had never flown a plane. He had no "kerosene in his blood." He was an MBA, a finance person. Before American, he'd worked at Hallmark Cards and Bloomingdale's.

Crandall, however, had a genius for finding creative ways to clean up a mess—for shaking out the contents of a purse, reorganizing it, making it more efficient. And he didn't care whom he pissed off while he did it. In other words, Crandall was a fire-breathing, gun-toting S-type innovator.

Trippe was a pilot who understood engines, loved flying, and designed planes like an engineer. After Trippe finished college, he collected a small pot of money from wealthy friends (his father had died and left little inheritance), bought a handful of war-surplus planes, and set up shop as Long Island Airways. It was Long Island, 1922—the summer that F. Scott Fitzgerald wrote about: Jay Gatsby and Daisy Buchanan, jazz and flappers. Rich couples would pay to fly out to the Island. Trippe's planes were among the best available, but they could carry only one pilot and one passenger. No couples. So Trippe modified his planes. He found a top-of-the-line French engine that could generate more power, cut down its oversize propellers, moved the fuel tanks to the outside of the fuselage, and added an extra seat. Business boomed.

For the next four decades, Trippe applied this strategy over and over. He designed and demanded bigger and faster planes that no one thought could be built, from his three-seater taxi all the way to the Boeing 747. Pan Am launched the Jet Age, brought international travel to the masses, and became the largest airline in the world. Trippe was a quietly dominant P-type innovator.

Do you remember that American Airlines created the frequent flier program? Or SuperSaver fares? Or do you have any idea what two-tier employment is? Unless you're an aviation industry historian, probably not. But if you're old enough, you probably remember Pan Am and the jet set. ABC ran a series about the lives of Pan Am pilots and stewardesses. No one will be making a TV series about airline reservation systems. Deregulation, however, created a special situation that briefly blocked the bright light of Juan Trippe–style changes and favored the dim glow of Bob Crandall–style changes. Deregulation created an S-type moment.

Let's understand one of those Bob Crandall changes a little bit better—the one change that Crandall, in an interview years later, described as the most critical to American's success. It was also the least sexy, least obvious change of all—so technical that we have to imagine a different business just to explain it. Ready?

THE PIE INDUSTRY: A BRIEF INTERLUDE

Let's say you have a business baking pies in Smalltown, USA. The laws of Smalltown, however, require pie-business owners to pay bakers $15 per hour. You and all the other owners sign contracts with the bakers agreeing to pay them $15 per hour for years, maybe decades. You're minding your own business when one day, the mayor of Smalltown decides, what the heck, you pie people are on your own, I'm finished with, um, sticking my fingers in your pies. Do whatever you want. The next day moms and pops all around Smalltown decide the pie business sounds splendid. New pie businesses open everywhere. They have no old contracts so they start paying their bakers $8 per hour! What do you do? Because their costs are lower, the new pie-business owners can afford to charge a lot less for their pies than you can. You can't lower your prices. You're locked into long-term contracts at insanely high rates. You're about to go out of business fast.

That's pretty much what the major airlines faced in 1978. They were locked into long-term contracts, paying wages far higher than what brand-new competitors were paying.

Crandall invented his way around the problem. He came up with the first two-tier pay system in American business. He created an A-scale, for employees hired before 1978, and a B-scale for new hires. He convinced extremely skeptical unions that two people doing the exact same job could be paid very different salaries. In exchange, the B-scale, market-rate costs allowed him to expand, buying more planes. The expansion meant more jobs and promotion opportunities, which the unions liked. The expansion lowered American's *average* labor costs to a breakeven point where the benefits of being big, with more reach, could make up for the startups' lower costs but smaller reach. It worked. American avoided bankruptcy, expanded, and eventually became the #1 airline in the country. There was no flashy new technology. Just a creative strategy on salaries.

Here's another one from the Bob Crandall loonshot nursery. Imagine Smalltown before Google and Yelp. How would residents know where to find the nearest, best pie-maker? The one that serves exactly the slice of pie they want? Yellow pages? Reaching out to each pie business with a device called a telephone? Inefficient.

So to help these hungry Smalltowners, as a conscientious provider of pies, you build a computer device that shows, on one screen, every pie shop in town—what they're serving, how much they're charging—and lets any user order any pie they want, anywhere. And, in your generosity, you give that pie-locator tool to every household in town for *free*. But wait! Other pie shop owners would *never* let you do that—because *of course* you'll just list your pie shop and no one else's. No, you say, I have to show *all* pie businesses, otherwise no one in Smalltown would accept the tool. Uh huh, say your competitors. But you win. Soon most of Smalltown is using your tool.

You've promised to be fair. Yet, mysteriously, your pie business goes up. Way up. And your competitors' pie sales go down. Way down. Who knew? Might it possibly have something to do with *placement*—with your pies being listed at the top of every screen?

Although American was not the first to develop a computerized reservation system, it developed the most functional one, which listed all fares, and then gave that system, Sabre, to travel agents all over the country. One study showed that American got *at least* 50 percent more business from travel agents who used Sabre than from other bookings. In an industry where one percentage point can make the difference between meeting payroll and bankruptcy, this really matters.

Crandall described the airline business as "the closest thing there is to legalized warfare." One competitor described Crandall's strategy as "cannibalistic. His goal is to kill the weak." Sabre gave American an immense battlefield advantage.

But the most crucial advantage from Sabre surprised Crandall and his team. They were soon flooded with data no one had ever seen before: years of bookings, as one analyst noted, "from which American could deduce how many days in advance vacationers tended to book to San Juan, how many days in advance business travelers booked to Detroit, in May as opposed to September, on Tuesday as opposed to Friday." Thirty years before Big Data became a Silicon Valley buzzword, American discovered big data. Crandall set up a division to use that data to extract maximum dollars per seat. The technique, as expected, was given a very boring name: yield management.

The frequent flier program that American invented around this time, which built customer loyalty, and the SuperSaver program, which filled empty seats with last-minute bookings, were much more visible. Other airlines quickly copied those ideas. But the behind-the-scenes, unglamorous, locked-in distribution channel from Sabre and the yield-management techniques from Big Data were almost impossible to copy, for many years. Those changes saved American.

Few people go wild for a reservations system. Most people have eyes only for the more glamorous types of loonshots. Including polite Yale gangsters named Juan Terry Trippe.

JT AND LINDY

A few years after designing his single-engine, three-seater Long Island air taxi, Trippe founded a new company. He worked with a Dutch aircraft designer named Anthony Fokker to custom-build a three-engine, eight-seater plane for that company. On January 16, 1928, flying the Fokker F-VIIa/3m, Pan American Airways launched its first passenger service, flying from Key West, Florida, to Havana, Cuba. Trippe's brochure advertised wide wicker chairs and sliding glass windows and asked, "How many times have you stood on the deck of a steamer, tossing in a rough sea? . . . How you longed for the smooth, quick flight of the gulls."

Business grew, but slowly. Trippe needed some way to fire up the public about flying. Which is when Trippe got the luckiest break of his career. He met and recruited Lucky Lindy.

At 7:51 a.m. on May 20, 1927, at Roosevelt Field, Long Island, Charles Lindbergh, age 25, eased back the stick on his single-engine plane, revving the engine, sending the heavily loaded plane wobbling down the runway. His plane, the *Spirit of St. Louis*, had never flown with this much weight. By the abort point on the runway, just past the halfway mark, the plane had not yet reached flying speed, but Lindbergh felt "the load shifting from wheels to wings." He cleared the telephone wires at the edge of the field by twenty feet.

So began Lindbergh's legendary quest to claim the $25,000 prize for the first nonstop flight between New York and Paris. He was the only one to attempt the flight solo, the only one in a single-engine plane. He took five sandwiches, a jug of water, and no navigation equipment other than a compass and a map. Lloyd's of London would not offer odds on his flight because the risk was too great. Eighteen people died in ocean-crossing attempts that year.

Lindbergh captured the public's imagination in a way that is hard to imagine today. The more he turned down interviews, the more his fame grew. The day before his flight, a cameraman had asked Lindbergh's mother to give him a kiss for a photo. She declined, smiling. "I wouldn't mind if we were used to that, but we come of an undemonstrative Nordic race." Reporters dropped any pretense of neutrality. "A slim, tall, bashful, smiling American boy is somewhere over the middle of the Atlantic Ocean, where no lone human being has ever ventured before," one columnist wrote. "If he is lost it will be the most universally regretted loss we ever had." The night Lindbergh's plane flew over the Atlantic, 40,000 boxing fans, gathered for a fight in Yankee Stadium, took a moment to offer a silent prayer. "Not even Columbus sailed alone," wrote one biographer. "Practically everybody who lived in America through Lindbergh's flight would remember his or her precise feelings that first night."

US newspapers ran over 250,000 stories about the flight. When Lindbergh landed outside Paris 33.5 hours later, a crowd of about 150,000 people rushed him and his plane, nearly tearing both apart in a tidal wave of adulation. Thirty million people, over one-quarter of the US population, came out to see Lucky Lindy during his three-month, 82-city tour to promote aviation. He was bombarded with job offers and sponsorship requests, from movie studios to shaving creams. Lindbergh resisted, saying he wanted to focus on promoting aviation.

The US ambassador to Mexico, Dwight Morrow, invited Lindbergh on a goodwill tour of Latin America to help repair strained political relations. Lindbergh, eager to escape a stream of society invitations, accepted.

The day he arrived in Mexico City, the ambassador's daughter, Anne, wrote in her diary, "I saw standing against the great stone pillar, a tall, slim boy in evening dress—so much slimmer, so much taller, so much more poised than I expected." They were married 18 months later.

On the goodwill tour, during a stop in Havana, Lindbergh connected with another young, patriotic American pilot with grand visions for international aviation. Lindbergh had turned down all the movie studios and shaving cream companies. But he accepted an offer from Juan Terry Trippe. He agreed to work with Trippe to build and promote Pan American Airways. The relationship would last four decades and change both of their lives.

Lindbergh began by advising Trippe on which planes to choose for flights to Latin America. He recommended an amphibian design, because of the region's lack of ground runways. Working with Igor Sikorsky, a Russian-émigré engineer, Lindbergh designed a custom aircraft: the S-38 "flying boat." The plane helped expand Pan Am's reach from two airports to over three dozen cities and ports across Central America, South America, and the Caribbean.

But where Lindbergh helped Trippe the most was in Washington. The

Charles Lindbergh and Juan Trippe plan Pan Am's conquest of Latin America (1929)

US Postal Service at the time hired private contractors for mail runs. At Trippe's request, Lindbergh lobbied on behalf of Pan Am for its Latin America routes. Imagine you are a career Post Office bureaucrat, and the most worshipped young man on the planet walks into your drab, ten-by-ten-foot office. Pan Am won *every* US postal contract to the region. Three round-trip mail runs to Buenos Aires alone paid for one Sikorsky S-38 flying boat. Without the airmail contracts, Trippe's competitors—other startup airlines eyeing the same routes—folded.

Trippe now had the routes and the planes. There was just one problem: navigation.

Trippe's pilots, like Lindbergh, flew by dead reckoning—compass, map, and eyeballs. Although the flight across the Straits of Florida was much shorter than Lindbergh's transatlantic flight, it was still dangerous. The islands and Florida Keys were much smaller targets than the shoreline of Europe, and the commercial planes had much shorter range than Lindbergh's single-engine, stripped-down plane. In his memoir, Lindbergh described a nighttime flight from Havana in early 1928:

> Over the Straits of Florida my magnetic compass rotated without stopping. . . . I had no notion whether I was flying north, south, east, or west. A few stars directly overhead were dimly visible through haze, but they formed no constellation I could recognize.
>
> I started climbing toward the clear sky that had to exist somewhere above me. If I could see Polaris, that northern point of light, I could navigate by it with reasonable accuracy. But haze thickened as my altitude increased. High thin clouds crept in to make the stars blink.

Lindbergh circled until dawn. In the early light, he checked his maps and discovered "I had flown at almost a right angle to my proper heading . . . it put me close to three hundred miles off route!" He was over the Bahamas, rather than Florida. Lindbergh survived because of the extra gas packed into the *Spirit of St. Louis.* (It was the penultimate flight of his famous plane. Two months later, Lindbergh flew it to Washington, D.C., and donated it to the Smithsonian, where it can be seen today.) Unlike the *Spirit*, however, the S-38 had a maximum range of 600 miles. A pilot flying 300 miles off course over water would be in serious danger.

At 3:55 p.m. on August 15, eight months after Pan Am's first passenger flight, Pan Am's recently recruited third pilot, a 33-year-old former Army pilot named Robert Fatt, departed Havana for Key West with two passengers and a navigator. Fatt had a grand total of four hours and twenty minutes of experience on a multiengine plane. The plane's radio receiver was in the shop, so Fatt could only transmit. One hour in, Fatt radioed to Key West: visibility was poor. It was raining. They were looking for signs of land but hadn't spotted any. Not worried. Another hour—still no visibility. And no land. Not worried.

Just under three hours later, Fatt made his final transmission, then crashed into the ocean. He was roughly 300 miles off course. Fortunately, a tanker was nearby. Three were rescued, including Fatt and his navigator. The fourth, one of the passengers, disappeared into the sea. The crash and the death shook confidence around the world not only in Pan Am, but in the future of commercial aviation. Trippe, somehow, managed to smooth-talk his way out of the mess. But he knew he had to solve the navigation problem, and quickly.

For answers, Trippe turned to a new idea: navigating by radio. Planes flying in poor visibility would transmit a signal to an operator on the ground, who would somehow decode their position and then radio back real-time instructions on their course. Pilots hated the idea, not wanting to trust control of their plane to someone miles away. And in any case, the only radio equipment at the time, used by ships, was unreliable and weighed hundreds of pounds—far too heavy for a plane.

People say, "There's no way that could ever work." And then it does.

Trippe invited Hugo Leuteritz, a radio engineer at RCA, the electronics company, to quit his job and join him full-time at Pan Am. Trippe knew that Leuteritz had proposed a portable radio navigator for planes, but RCA had rejected the idea. Leuteritz, 31, was intrigued by Trippe's offer but had a steady job and a young family.

"You have only a few airplanes," Leuteritz told Trippe. "That's not enough to keep me busy."

"We will have a big fleet of planes," Trippe answered. "We will fly to Latin America next year, and after that we will fly across the Atlantic, across the Pacific." (Trippe was all of 28 and had been in business barely one year.)

Leuteritz left his company and joined Trippe. He was shown his new office: a chair and his lap. Within a year, Leuteritz gave Trippe the final piece to his puzzle. By the end of 1929, Pan Am had 25 ground-based radio receivers directing its fleet of 60 planes, each fitted with an industry-first, lightweight radio-navigation system. That year Pan Am flew 20,728 passengers 2,752,880 miles between 60 airports in 28 countries. There were no more planes lost at sea. And Pan Am had become the largest international airline in the world.

Now Trippe added one more element: glamour. In the fall of 1929, Lindbergh and his new wife, Anne, toured Latin America with Trippe and his wife, Betty, in an S-38. The women wore white and served tea. The men wore flight goggles and posed for pictures with politicians. Cheering crowds greeted the two twentysomething couples at every city. The spectacle grew as Anne Lindbergh's occasional airsickness was revealed to be morning sickness. The public followed the pregnancy like the arrival of a royal heir.

THE DANGEROUS VIRTUOUS CYCLE

Trippe's strategy of nurturing P-type loonshots and betting on bigger, faster, more—with a dash of marketing glamour—worked brilliantly. Technology improvements lowered costs, providing more money to invest in more technology improvements. Larger planes flew more customers farther, faster. That virtuous cycle continuously grew his franchise, propelling Trippe far ahead of competitors, attracting fame and celebrity, just as a similar virtuous cycle would propel leading technology companies for the rest of the century, from Polaroid to IBM and Apple. P-type loonshots feed a growing franchise, which feeds more P-type loonshots. And as the momentum builds, so does the tunnel vision: keep turning the wheel, faster and faster.

With each turn of the cycle, as Pan Am's planes and influence grew, so did Trippe's ambition. Traveling between the New World and Old would no longer be only for elites: Pan Am would offer affordable, same-day travel across the oceans for everyone.

Trippe began with plans to cross the Atlantic, the most lucrative air route. A million passengers and 75 million pounds of cargo traveled the 10-day ocean voyage by steamship each year. The volume dwarfed Pan Am's Latin America business. But after four years of negotiations for European air and landing rights ended in failed agreements, Trippe shifted his attention to the Pacific. The goal of commercial flights across the Atlantic was considered ambitious and challenging. The idea of flying across the Pacific, on the other hand, was considered suicidal. The world's longest air route at the time was an 1,865-mile airmail flight between Africa and Brazil. Trippe proposed to fly passengers—not just mail—8,700 miles over poorly charted oceans. In a widely publicized competition for the first nonstop flight just to Hawaii only a few years earlier, three of the six planes to attempt the flight had been lost at sea. And getting to Hawaii was the easy part. The span between Honolulu and China, twice the width of the Atlantic, had no known bases for refueling. Trippe's longest-range plane, the Sikorsky S-42, could fly, fully loaded, less than one-fifth that distance.

When Trippe announced publicly that Pan Am would fly to China, two members of the Pan Am board of directors resigned, convinced that Trippe was leading the airline to tragedy. The chairman of the federal committee overseeing aviation, a friend of Trippe's, offered to have the government publicly object on the grounds of safety so Trippe could back down and save face. Trippe rejected his offer.

In the face of insurmountable odds, public opposition, and the threat of a national disaster, where do you begin? The New York Public Library, of course. Trippe went to the main branch on 42nd Street and Fifth Avenue. At the information desk, he asked for the logs of the nineteenth-century clipper ships that traded across the Pacific. Buried in the old handwritten documents, Trippe found a reference to a deserted island midway between Honolulu and Shanghai, called Wake Island. An American expedition had claimed the island in 1899. Trippe asked around in Washington; no one knew who administered it. A few phone calls and letters from Trippe resulted in a presidential executive order. Within months, the Navy was assisting Pan Am in building an air base on Wake Island and, soon after, on two other deserted American territories west of Hawaii: Midway and Guam. The three islands completed the stepping stones for a viable flight path between mainland US and Asia. Later, they would play a critical role during World War II.

Next, Trippe commissioned "a high-speed multimotored flying boat having a cruising range of 2,500 miles against 30-mile headwinds" from the Glen Martin Company of Baltimore, Maryland. On November 11, 1935, the world's largest amphibious plane—25 tons, 130-foot wingspan, four 830-horsepower Pratt & Whitney engines—glided into San Francisco Bay, fresh from Baltimore. The gleaming silver and blue Martin M-130 docked at the Alameda pier across from Oakland, bobbing in the waves. Trippe had named it the *China Clipper.*

The governor of California declared November 22 Pan American Airways Day. At 2:45 p.m., a radio announcer described the tens of thousands of people gathered along the shore, with millions more listening on the radio, "about to witness with us one of the most dramatic events in the history of our modern world." The postmaster general read a cable from President Roosevelt and declared that the day "will forever mark a new chapter in the glorious history of our nation, a new era in world transportation, a new and binding bond that will link, for the first time in history, the peoples of the East and the West." As each of the seven crew members emerged onto the pier and walked toward the plane, like baseball players emerging from a dugout, the announcer called out their names and bios to cheers and applause.

The crew boarded and the doors locked. One by one, air stations for each stop along the way radioed in: Honolulu, Midway, Wake, Guam, Manila—standing by. Trippe reported all stations ready; the postmaster general gave the order to proceed; the *Clipper*'s engines roared; "The Star-Spangled Banner" blasted; hundreds of car drivers honked; and 22 aerial bombs exploded in salute. Thirty small, circling planes descended, tiny consorts to accompany the inaugural flight of the royal queen bee.

The *Clipper* rose—but, loaded with nearly two tons of airmail, it struggled to climb. With growing speed, it headed straight toward the cables of the unfinished San Francisco–Oakland Bay Bridge. At the last moment, the pilot abandoned the plan to fly over the bridge, dropped the nose, and flew underneath. "We all ducked," said one of the engineering officers later. The announcer and the crowd believed this was part of the show and cheered. Back on the pier, Trippe knew better, and flinched. Emerging from under the Bay Bridge, the plane found lift, climbed over the Golden Gate Bridge (also under construction), and then settled in for the 21-hour flight to Honolulu. The remainder of the trip was uneventful.

By the time the *Clipper* returned, one week later, the captain, Ed Musick, was—after Lindbergh—the most famous pilot in the country, and Pan Am was the most famous airline. Magazines printed glamorous photos of the white-tablecloth dining cabin; stewards in tuxedos served passengers like Ernest Hemingway; and Warner Brothers released a movie, *China Clipper*, about a young entrepreneur who builds a plane to fly across

The *China Clipper* ducks underneath the San Francisco Bay Bridge (1935)

the Pacific. The fourth-billed role of the captain was played by a 37-year-old B-movie actor named Humphrey Bogart.

The cycle of cash pouring in from a growing franchise, funding those P-type loonshots that further feed the franchise, was revving at a high pitch. That momentum would continue to build for another 20 remarkable years, leading up to Pan Am's most spectacular success, the peak glory of Trippe's vision.

WARS, LOONSHOTS, AND CUCKOO CLOCKS

> In Italy, for 30 years under the Borgias, they had warfare, terror, murder and bloodshed, but they produced Michelangelo, Leonardo da Vinci, and the Renaissance.
>
> In Switzerland, they had brotherly love, they had 500 years of democracy and peace—and what did that produce? The cuckoo clock.
>
> —Orson Welles as Harry Lime, *The Third Man* (1949)

In May 1939, Pan Am finally crossed the Atlantic. Only four months of peacetime travel remained, however. Hitler invaded Poland in September of that year. Trippe was asked to assume active duty as an Air Force general. He declined, but Pan Am was drawn into the war nonetheless.

Trippe would soon discover what Harry Lime—the character played by Orson Welles who thrived off the chaos of postwar Vienna—so patiently explained to his more innocent friend, Holly Martins: wars accelerate P-type loonshots.

In June 1940, FDR asked Trippe to build 25 new airports in South America, which could double as American bases, under the cover of Pan Am's commercial needs. America was officially neutral, but FDR knew of the strong German ties to the continent. Trippe agreed. One year later in London, at a private dinner at the prime minister's residence, Churchill asked Trippe to set up an air route to resupply British troops trapped in Africa. Again, Trippe agreed.

But perhaps the strangest request came at a secret meeting in China, just after the war, when Trippe was passing through Shanghai, a stop during the first commercial round-the-world flight. As he was relaxing in his hotel room after dinner, he heard a knock on the door. He opened it to find China's finance minister, Chang Kia-ngau, standing there. Chang apologized for the late interruption and told Trippe that in the morning there would be a taxi with a red O on the windshield waiting for him downstairs. The driver would bring Trippe to the back door of a private residence. Once there, Chang would present Trippe with "the plan." He bowed and left.

Trippe thought it might be a joke. But the next morning, there was a taxi with a red O waiting for him. He got in and was driven to the private home, where he was led through a basement and a garden and delivered to Chang. The minister explained to Trippe that the country's leader, Generalissimo Chiang Kai-shek, would like America to save China from the Communists. America would appoint a high commissioner to rule China, as it had in Japan, and train Chiang's Nationalist army to battle Mao's Communists. Mao's forces were small but growing quickly, funded by the Soviet Union.

Chiang asked Trippe to deliver the plan, verbally, to President Truman. He explained that he believed Trippe would have more influence than the American ambassador. Trippe, in disbelief, asked for the plan in writing. Signed.

The next morning, the finance minister surprised Trippe at his hotel once again. He offered to accompany him to the airfield where Trippe's plane was waiting. When they arrived, Chang asked Trippe to show him

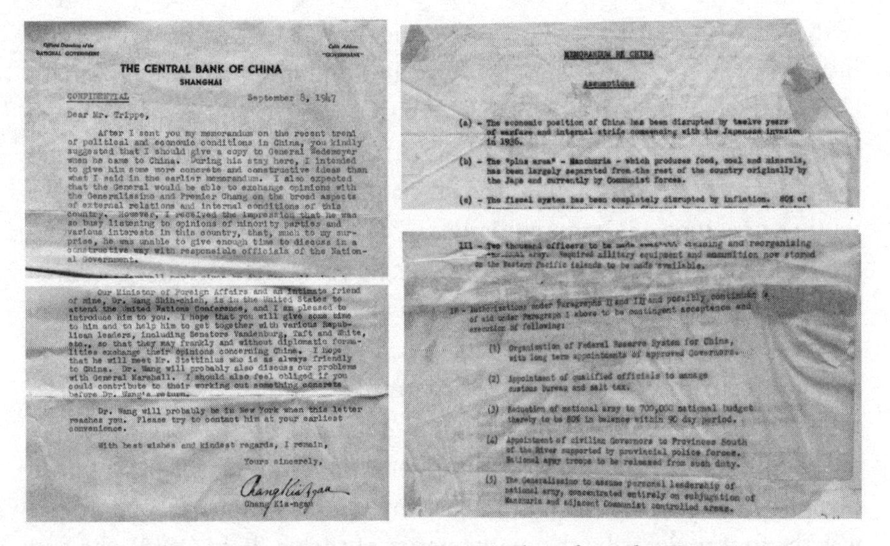

THE CENTRAL BANK OF CHINA
SHANGHAI

CONFIDENTIAL September 8, 1947

Dear Mr. Trippe,

After I sent you my memorandum on the recent trend of political and economic conditions in China, you kindly suggested that I should give a copy to General Wedemeyer when he came to China. During his stay here, I intended to give him some more concrete and constructive ideas than what I said in the earlier memorandum. I also expected that the General would be able to exchange opinions with the Generalissimo and Premier Chang on the broad aspects of external relations and internal conditions of this country. However, I received the impression that he was so busy listening to opinions of minority parties and various interests in this country, that, much to my surprise, he was unable to give enough time to discuss in a constructive way with responsible officials of the National Government.

Our Minister of Foreign Affairs and an intimate friend of mine, Dr. Wang Shih-chieh, is in the United States to attend the United Nations Conference, and I am pleased to introduce him to you. I hope that you will give some time to him and to help him to get together with various Republican leaders, including Senators Vandenburg, Taft and White, etc., so that they may frankly and without diplomatic formalities exchange their opinions concerning China. I hope that he will meet Mr. Stettinius who is as always friendly to China. Dr. Wang will probably also discuss our problems with General Marshall. I should also feel obliged if you could contribute to their working out something concrete before Dr. Wang's return.

Dr. Wang will probably be in New York when this letter reaches you. Please try to contact him at your earliest convenience.

With best wishes and kindest regards, I remain,

Yours sincerely,

Chang Kia-ngau

MEMORANDUM RE CHINA

Assumptions

(a) - The economic position of China has been disrupted by twelve years of warfare and internal strife commencing with the Japanese invasion in 1936.

(b) - The "plus area" - Manchuria - which produces food, coal and minerals, has been largely separated from the rest of the country originally by the Japs and currently by Communist forces.

(c) - The fiscal system has been completely disrupted by inflation. 80% of

III - Two thousand officers to be made available for training and reorganizing national army. Required military equipment and ammunition now stored on the Western Pacific islands to be made available.

IV - Authorizations under Paragraphs II and III and possibly continuing of aid under Paragraph I above to be contingent acceptance and execution of following:

(1) Organization of Federal Reserve System for China, with long term appointments of approved Governors.

(2) Appointment of qualified officials to manage customs bureau and salt tax.

(3) Reduction of national army to 700,000 national budget thereby to be 80% in balance within 90 day period.

(4) Appointment of civilian Governors to Provinces South of the River supported by provincial police forces. National army troops to be released from such duty.

(5) The Generalissimo to assume personal leadership of national army, concentrated entirely on subjugation of Manchuria and adjacent Communist controlled areas.

Chang Kia-ngau asks Juan Trippe to save China from the Communists
(excerpts of letter from Chang to Trippe, Sept. 8, 1947)

the inside of the plane. Trippe agreed. Chang seemed uninterested in the tour, however, and whispered to Trippe to join him in the tiny bathroom. Standing inches from the six-foot-tall, bulky, and suddenly quite uncomfortable Trippe, the finance minister pulled out the written plan. He pointed at Chiang Kai-shek's signature and asked Trippe to deliver it personally to Truman. Trippe stared at the Chinese signature. He asked Chang to sign it as well. Neither had a pen, so Trippe went to the cockpit, borrowed a pen, and returned to the lavatory. Chang signed the document on the tiny sink. When he returned to Washington, Trippe duly passed the plan on to Truman.

Trippe's journey through film noir in Asia did not change the fate of China or Pan Am in the end. It was the rise of a dangerous new loonshot in Europe, toward the end of the war, that would change not only Trippe's life, but world travel, more than any invention since the airplane.

* * *

On July 25, 1944, a Royal Air Force pilot flying a British DH-98 Mosquito over Munich observed a new kind of German plane streaking toward him. It had no propellers and flew 120 miles per hour faster than any British or American aircraft. After a close dogfight, he managed to escape into cloud cover.

Seven weeks later, Bernard Browning, age 28, a British Army engineer, was walking down Staveley Road in London on the way to see his girlfriend when the street exploded. All that was left where he had stood was a 30-foot crater. British officials attributed the explosion to a gas main. Reporters didn't believe the official explanation. They guessed the truth: a new type of German missile.

Two centuries after the steam engine, 80 years after the first gasoline-fueled combustion engine, the Germans had unlocked a new source of power—the jet engine. The RAF pilot, A. E. Wall, had witnessed the first jet-powered plane, called the Messerschmitt Me 262. Browning had been killed by the first ballistic missile, the V-2 rocket.

The principles and design behind jet engines and rockets had been established 25 years earlier by an American physicist, Robert Goddard. He was the first to describe the mathematics of rocket flight (1912), the first to design and build a liquid-fueled rocket (1926), and the first to demonstrate gyroscopic stabilization of rockets (1932). His ideas were dismissed by academics and the military in the United States. A *New York Times* editorial stated that Goddard "seems to lack the knowledge ladled out daily in high schools," namely that of Newton's law on action and reaction, which made rocket flight impossible. (Forty-nine years later, the day after the successful launch of the Apollo 11 rocket to the moon, the paper issued a retraction. It announced that rockets did not in fact violate the laws of physics and that "the *Times* regrets the error.")

Scientists in Germany, however, had taken Goddard's ideas seriously. They began their program after reading his papers. Years later, a German officer being grilled by American officers about their V-2 rocket program exclaimed, "Why don't you ask your own Dr. Goddard?"

In the US, one well-known aviation expert had also taken the idea of rocket propulsion seriously. Charles Lindbergh had encouraged Goddard and introduced him to the donors who funded his research. Lindbergh was a full colonel in the Air Force. But he was unable to interest the military in Goddard's rockets. Much of that was due to an unprecedented campaign of personal attacks against Lindbergh by President Roosevelt.

A conflict between the two public giants had been brewing for years before the war. The conflict began when Roosevelt canceled postal service airmail delivery contracts in favor of having the Army Air Corps deliver the mail. Lindbergh strongly and publicly opposed the move, noting the dangerous lack of army experience in flying in the extreme weather conditions on many mail routes. Air Corps pilots crashed 66 times, resulting in 12 deaths. Roosevelt was forced to reverse his decision, a humiliation that made front-page news. The fight "dented the myth of Roosevelt's invulnerability," wrote one historian, and "uncovered in Charles Lindbergh a man who perhaps appealed to more American hearts than anyone save Franklin Roosevelt."

In 1939, Lindbergh began publicly opposing any US intervention in Europe. He spoke at massive antiwar rallies and attacked Roosevelt. Crowds flocked to Lindbergh, shouting, "Our next president!" Roosevelt, who had a long memory, began a campaign to undermine Lindbergh. Publicly, he called Lindbergh a "defeatist and appeaser." Privately, he vowed, "I'll clip that young man's wings." Lindbergh was soon shunned: the press crucified him as a Nazi sympathizer and traitor; streets that had been named in his honor were renamed. One city threatened to burn Lindbergh's books in a public square.

"In just fifteen years," Lindbergh's sister said, "he went from Jesus to Judas." Lindbergh had no influence left with the military to recommend Goddard's rockets or any technology.

Lindbergh also found himself looking for a job for the first time in 20 years. Although he had not worked officially for Trippe and Pan Am for many years, he reached out to Trippe and was warmly received, offered any position he wanted. A few days later Trippe called and withdrew the offer. The White House, Trippe explained, had angrily insisted Pan Am have no connection with Lindbergh.

In April 1945, a month before Germany surrendered, FDR died. Resistance to Lindbergh in official circles disappeared, and the Navy called him to Washington. There were rumors of new kinds of planes and missiles in Germany. Something like the rockets he had mentioned six years earlier. Would Lindbergh join a secret mission to Germany to investigate?

Lindbergh quickly received State Department clearance for a secret Naval Technical Mission. In Europe, Lindbergh tracked down Willy Messerschmitt, who shared details of his famous plane. At the BMW factory that produced jet engines, a German engineer who looked "a little white and shaky" approached Lindbergh. Before American troops had arrived, he said, he had been given drawings for one of the jet engines with orders to destroy them. Instead, he had buried them in a file box underneath a large pine tree a short drive away. Would Lindbergh like the files? The two drove to the tree, parked, and began digging. Soon the shovel hit a metal box. Lindbergh had the designs to Germany's jet engines.

When Lindbergh returned to the US, he filed his report, then immediately called on Trippe. Trippe rehired Lindbergh on the spot. It was time for the next turn of the loonshot-franchise cycle.

It was time to build a new kind of plane.

THE JET AGE

Trippe quickly began discussions with Boeing, which was then primarily a builder of military aircraft but looking for a way in to the commercial market dominated by its competitors, Lockheed and Douglas. Boeing told Trippe that it would build Pan Am a commercial jet if he would place a firm order. But the range Boeing proposed was too short for Trippe, and the fuel consumption too high. Trippe passed.

Lindbergh and Trippe discovered that the UK was far ahead of the US in building jets. The British national airline (BOAC) had already commissioned a commercial jet from de Havilland Aircraft, a British manufacturer. In 1952, the de Havilland Comet began service, to triumphant British flag-waving. The chairman of BOAC declared, "This present Elizabethan age is repeating in the air what the first Elizabethan era saw at sea." The celebration lasted less than two years. In 1953 and 1954, three unexplained midair explosions of Comets killed all passengers on board. The government grounded the entire fleet.

The Comet explosions scared most of the industry off jets. In case there were any more doubts, a report by the Rand Corporation, the premier national policy and security consulting firm, said that jet travel could never be made economically feasible (BOAC jets flew at a loss). The presidents of American Airlines and Trans World Airlines (TWA) announced they would not pursue jet aircraft.

People say, "There's no way that could ever work." And then it does.

After countless discussions with engineers and a review of the Comet data, Trippe concluded that the safety and financial problems of the Comet jets were fixable. The Comet explosions were tragic failures. But in the language of the previous chapter, they were a False Fail. Just as Thiel and

Howery dug into the demise of Friendster and came up with a contrarian view on social networks, Trippe and Lindbergh came up with a contrarian view on jet airliners.

A British government investigation eventually confirmed Trippe's view on safety: metal fatigue caused by a unique window design had caused the accidents. Metal fatigue was a fixable engineering problem. The economics could be addressed by a different plane design: the Comet's range was too short, its passenger capacity (44 seats) too low, and its fuel consumption too high.

Trippe went back to Boeing, which by then had developed a prototype commercial jet called the 707. He immediately saw that the 707 had the same drawbacks as the Comet. Trippe wanted a jet that could fly transatlantic, nonstop. He politely insisted on a redesign. Boeing wouldn't budge. It had already invested millions building the plane and wouldn't scrap it.

So Trippe and his team went to Santa Monica, to convince Boeing's chief competitor, Donald Douglas, to build the jet Trippe wanted. Douglas also declined. All the other airlines were ordering his market-leading propeller plane, the DC-7, and Douglas saw no reason to build a jet. Trippe persisted and eventually convinced Douglas to submit a proposal. The design was similar to Boeing's. Not good enough. The problem, Trippe realized, was the engine: the industry-best Pratt and Whitney J-57 could not support a nonstop transatlantic flight. So Trippe sent his team, including Lindbergh, to Pratt and Whitney.

At the engine manufacturer, they learned about an experimental engine with a new high-compression technology that produced up to 50 percent more power and drastically improved fuel efficiency. It could do the job. But it was far from ready and still on the military secret list.

Trippe asked Frederick Rentschler, the founder and president, to release the new engine. He, Trippe, would get it cleared with the military. The answer was a resounding no. The engine was too experimental, and Rentschler had plenty of other orders for his current engines.

And this is where Trippe performed a minor business miracle, the last and possibly greatest of his career. Three of the most successful manufacturing firms in the world, led by legendary entrepreneurs and businessmen, had turned him down. He got them all to reverse their decisions.

First, Trippe began talking to the British engine manufacturer

Rolls-Royce. The British company was also working on a secret next-generation jet engine. Just as Trippe had planned, Rentschler at Pratt and Whitney heard about their discussions. Rentschler convened an emergency internal meeting. Could Pratt and Whitney afford to lose Pan Am's business? Could they accelerate development?

Shortly afterward, Trippe called. He had a new offer for Rentschler: Pan Am would buy his engines directly. An airline buying engines, without the planes to put them in? Exactly. And Trippe would like 120 of them, a $40 million order. (This was an enormous sum in the 1950s, four times Pan Am's annual income.) Rentschler made the final decision. Yes. Trippe could have his new engines.

Next, Trippe flew to Seattle. Would Boeing build the planes he wanted now that he owned the engines? If not, he would look elsewhere. The president, Bill Allen, called Trippe's bluff. The answer was still no. So Trippe flew to Santa Monica. Would Douglas build the plane now that Trippe had the engines? Douglas realized that with the engines, Trippe was all in and would likely find a builder somewhere for what might become the best plane in the world. He caved. Douglas would build a plane, the DC-8, to Trippe's design. Trippe committed to 25 planes, but asked Douglas to hold off on his announcement.

Trippe flew back to Seattle, met with the Boeing team, and agreed to their offer—he would buy 20 of their smaller planes, the 707, which could not make the transatlantic flight. He didn't tell them about the Douglas order. The Boeing team celebrated: they had succeeded in convincing the infamously stubborn Trippe to be reasonable.

Trippe timed the joint press releases. On October 14, 1955, Allen and Douglas opened the *Wall Street Journal* and learned about each other's orders. The Boeing president later said he "felt like an earthquake victim." After betting his company on a new plane, it was instantly obsolete. The message was clear to anyone reading the paper: 25 orders for Douglas's superior plane versus 20 for Boeing's inferior plane. Allen called Trippe and conceded. He could have his redesign. Boeing could not afford to be second-class.

The two manufacturers were now competing to build Trippe the best possible plane, to his design. Other airlines abandoned their contracts for propeller planes and rushed to order the new jets. Trippe had played possibly the highest-stakes game of business poker in corporate history up

Pan Am launches the Jet Age

to that time—a $269 million order for 45 unprecedented commercial jet planes—and won.

Both the Boeing 707 and Douglas DC-8, when they arrived, transformed travel. The average middle-class family, for the first time in history, could now afford convenient, same-day international or transcontinental travel.

Pan Am and Trippe rode the Jet Age, turning the franchise wheel faster and faster. By 1965, seven years after the first flight of the Boeing 707, traffic had grown by over 400 percent, net income by over 1,000 percent. Trippe added a hotel division, Intercontinental Hotels, and a business jets division. He commissioned the world's largest office building, on Park Avenue. The Air Force solicited bids for work on long-range missiles, so Trippe added a guided-missile division, and soon after an aerospace division, which supported the Apollo moon landing. Between 1968 and 1971, Pan Am accepted 93,005 reservations for a planned lunar service.

The franchise was growing, literally, to the moon.

ONE MORE TURN

And then, of course, Trippe heard about a new kind of engine. One more loonshot. A technology that *quadrupled* maximum takeoff weight. A plane with the new bypass jet engine, with its extra propeller in front, could carry nearly 500 passengers, two and a half times as many as the Boeing 707. The music of the Jazz Age, now the Jet Age, keeps playing. The wheel in the sky keeps on turning. The franchise feeds the P-type loonshots that feed the franchise. Bigger, faster, more. Trippe had to have it.

In August 1965, ten years after their first legendary deal, Trippe and Bill Allen, still the president of Boeing, went salmon fishing with their wives in Alaska. Trippe told him about the engine and the plane he wanted.

"If you build it," said Trippe, "I'll buy it."

"If you buy it, I'll build it," Allen responded.

One more dance. On December 22, they signed, once again, the largest corporate deal in history to that time: $525 million for 25 production models of a new, first-in-class plane. Allen gave it a name: the Boeing 747.

To fill two and a half times as many seats, an airline needs two and a half times as many passengers. Pan Am's lock on international travel, however, had been weakening. Congress had launched an antitrust investigation of Pan Am's monopoly on foreign routes in the 1950s. Populist voices complained that regulators were protecting industry giants rather than consumers. Some of the loudest complaints came from startups—Texas Air, Braniff, and, eventually, Southwest Airlines.

The startups also brought new ideas to the industry. Hub and spoke. Flying to secondary airports. Reducing turnaround times to 20 minutes. Like Sam Walton's supersized stores far outside cities, none of the ideas involved new technologies. They were all small changes in strategy that no one thought would amount to much. They were all S-type loonshots.

Boeing delivered its first 747 in January 1969. The music had stopped, but Pan Am didn't notice. It ordered eight more 747s for another $200 million. Then it splurged on a new $100 million terminal at Kennedy Airport in New York. While its competitors nurtured S-type loonshots, Pan Am doubled down on franchise. The faint glimmers from those S-type loonshots were dwarfed by the decade's most glamorous P-type loonshot: the 747 jumbo-jet airliner.

Until deregulation. Small changes that improved efficiency and lowered costs—not glamorous, kind of boring—suddenly became the key to survival. Those S-type loonshots, nurtured by startups like Southwest or major carriers like Bob Crandall's American, spread quickly through the industry. They annihilated every unprepared airline.

Pan Am began the steady decline from which it never recovered. After deregulation, Pan Am lost money every year for eight years. It stayed alive only by selling off pieces of itself. The office building in New York. The hotel business. The magical routes to China. The new terminal at Kennedy. Until, finally, there was nothing left to sell.

Trippe had retired, suddenly, in the spring of 1968, before the first 747 arrived. Maybe, at age 68, he was tired. Or maybe he lifted his gaze from engines and franchise and realized his theme song, from the movie he had starred in for 41 years, was no longer playing. He died in 1981. He witnessed the decline, but not the end.

A revolving door of CEOs came and went, but the organization had already hardened, it had already passed through its phase transition. It could only grow franchise, not nurture loonshots. One CEO spent $374 million to acquire a US domestic airline, National, adding dozens of new routes, trying to grow the franchise. Another CEO tried to buy Northwest Airlines for nearly $3 billion. It made no difference.

In December 1991, Pan American World Airways ceased to exist.

§

WATCH YOUR BLIND SIDE

The demise of Pan Am was a remarkable story, but not a unique one. Nearly every company led by a master P-type innovator like Trippe gets shocked. Some sudden change, whether from a regulatory agency or a new competitor, stops the music. The loonshot-franchise cycle stops working. The wheel turns one too many times, and suddenly there's a fleet of 747s that no one wants to fly. Competitors who have been nurturing their own loonshots, one or more of which fit the newly changed world, race by.

The S-type loonshots from American Airlines' Bob Crandall and others overtook Trippe on his blind side. The same can happen to entire teams, or companies.

The collapse, for example, of IBM's legendary 80-year-old hardware business in the 1990s sounds like a classic P-type story. New technology (personal computers) displaces old (mainframes) and wipes out incumbent (IBM). But it wasn't. IBM, unlike *all* its mainframe competitors, mastered the new technology. Within three years of launching its first PC, in 1981, IBM achieved $5 billion in sales and the #1 position, with everyone else either far behind or out of the business entirely (Apple, Tandy, Commodore, DEC, Honeywell, Sperry, etc.).

For decades, IBM dominated computers like Pan Am dominated international travel. Its $13 billion in sales in 1981 was more than its next seven competitors *combined* (the computer industry was referred to as "IBM and the Seven Dwarfs"). IBM jumped on the new PC like Trippe jumped on the new jet engines. IBM owned the computer world, so it outsourced two of the PC components, software and microprocessors, to two tiny companies: Microsoft and Intel.

Microsoft had all of 32 employees. Intel desperately needed a cash infusion to survive. IBM soon discovered, however, that individual buyers care more about exchanging files with friends than the brand of their box. And to exchange files easily, what matters is the software and the microprocessor inside that box, not the logo of the company that assembled the box. IBM missed an S-type shift—a change in what customers care about.

PC clones using Intel chips and Microsoft software drained IBM's market share. In 1993, IBM lost $8.1 billion, its largest-ever loss. That year it let go over 100,000 employees, the largest layoff in corporate history. Ten years later, IBM sold what was left of its PC business to Lenovo.

Today, the combined market value of Microsoft and Intel, the two tiny vendors IBM hired, is close to $1.5 trillion, more than ten times the value of IBM. IBM correctly anticipated a P-type loonshot and won the battle. But it missed a critical S-type loonshot, a software standard, and lost the war.

Learning to watch your blind side is one lesson, an important one. But there is a much bigger one. It's the key to the fourth quadrant identified at the end of chapter 1—the Trap.

For 41 years, Juan Trippe stood atop his mountain and anointed loonshots. He saw a new technology, a P-type loonshot, that might grow his franchise—a faster engine, a navigation system—and he had to have it. Even when no sensible strategy could justify bigger, faster, more—he had to have it.

Let's call it the *Moses Trap*: When ideas advance only at the pleasure of a holy leader—rather than the balanced exchange of ideas and feedback between soldiers in the field and creatives at the bench selecting loonshots on merit—that is exactly when teams and companies get trapped. The leader raises his staff and parts the seas to make way for the chosen loonshot. The dangerous virtuous cycle spins faster and faster: loonshot feeds franchise feeds bigger, faster, more. The all-powerful leader begins acting for love of loonshots rather than strength of strategy. And then the wheel turns one too many times.

The leader and his followers may reach for the moon, like Pan Am, and have their wings clipped. Or they may reach even higher, as was the case with our next Moses.

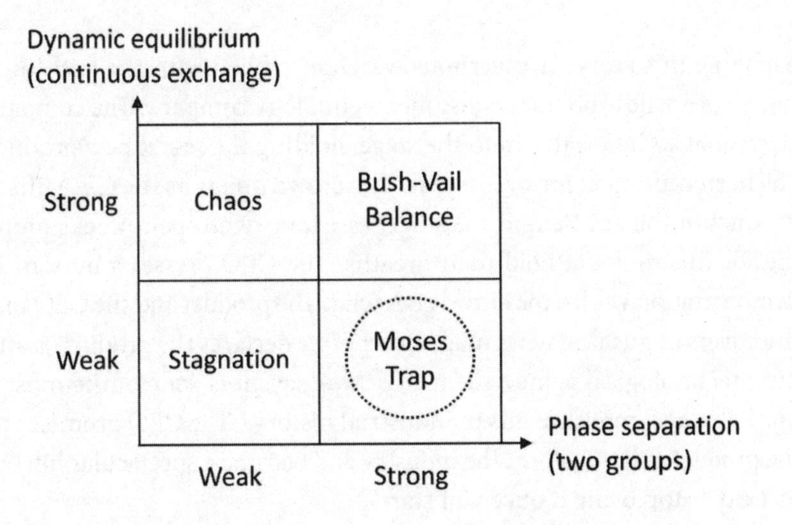

4

Edwin Land and the Moses Trap

When leaders anoint the holy loonshot

Imagine this scene: A cavernous warehouse filled with the faithful followers of a wildly popular consumer technology company. The company's charismatic CEO walks onto the stage, holding the secret new product it has been hinting at for over a year. The crowd quiets as the CEO lifts the product in the air. Behind the stage, assistants who spent weeks preparing for this moment hold their breaths. The CEO presses a button. The demonstration works, the crowd goes wild. The product and the CEO make the covers of gushing news magazines: *Time* declares the product "a stunning technological achievement"; *Fortune* says it is "one of the most remarkable accomplishments in industrial history." The CEO promises that the product will transform the industry and become a spectacular hit: "You just can't stop using it once you start!"

That must be Steve Jobs introducing the iPhone, right? It's not. It's Edwin Land, introducing the Polaroid SX-70—their iconic, pyramid-shaped, collapsible, instant-color-print camera—35 years earlier, in 1972. For 30 years, Polaroid scientists produced one Nobel-caliber breakthrough after another. They created new molecules, unlike anything seen before, that achieved the impossible—instant color prints. They invented a new theory of color vision that changed our understanding of the brain. They solved the century-old problem of separating light into its components, technology used in every smartphone display and computer monitor. The company was the glamour stock of its day, reaching new highs every year as rabid fans bought and bought.

And then something changed. The magic faded. Polaroid declined, descended into debt, and eventually filed for bankruptcy.

Juan Trippe began with a small airline taxi service and built a large airline empire. Edwin Land began with a hidden property of light and built an empire famous for something completely different. Both empires followed similar cycles to similar ends. Loonshots fed a growing franchise, which in turn fed more loonshots.

But as recently declassified documents show, Land led another life. That life sheds new light on the trap at the end of the cycle—and how to escape it.

HAN SOLO'S ESCAPE

A beam of light has three familiar properties: direction, intensity, and color. It also has a hidden fourth property, called polarization. Imagine a drone flying level to the ground. The drone can have wings parallel to the ground, rotated 90 degrees, or at any angle in between. Polarization of light acts like the wings on the drone. A light beam traveling parallel to the floor can be polarized horizontally, vertically, or at any angle in between. Our eyes can't detect polarization, so we don't see it.

Although the name of his company eventually became synonymous with something else, Edwin Land built the Polaroid Corporation by inventing remarkable uses for this hidden property of light.

If you are a *Star Wars* fan, you might remember the asteroid scene in *The Empire Strikes Back* (1980). TIE fighters are chasing the *Millennium Falcon*, piloted by Han Solo with Chewbacca and Leia at his side. Han steers into an asteroid field ("Never tell me the odds!"), plunges deep into a big cave on an asteroid, and lands the ship, waiting for the TIE fighters to pass by. The three step out to look around. They quickly realize the "cave" is not quite what they thought. They race back to the *Falcon*, fire it up, and fly at full speed toward the rapidly closing heavily fanged jaws of the giant worm (technically, an exogorth), in whose mouth they'd parked. The *Falcon* is horizontal. The worm's teeth are vertical. At the last second, Han flips the ship 90 degrees and escapes through the narrow slits between the teeth. The jaws snap shut behind him.

Polarizing filters function like the worm's teeth: a vertical filter only lets through vertically polarized light. The *Falcon* vertical passes through. The *Falcon* horizontal does not.

Land had wanted to make his own polarizer since age 13, when as a summer camp counselor he'd used a block of Icelandic crystal (a natural polarizer) to make the glare from a tabletop disappear. For a century, people had attempted to create a practical polarizer to unlock the mysteries of light, but no one had succeeded. Years later, Land became known for a saying: "Do not undertake a program unless the goal is manifestly important and its achievement nearly impossible." He began that summer. He slept with a book called *Physical Optics* underneath his pillow. He read the book "nightly in the way that our forefathers read the Bible."

At age 17, Land enrolled at Harvard. A few months later he left, bored of being surrounded by wealthy kids with no ambition. Land moved to New York City and convinced his skeptical father to continue his college allowance while he pursued his dream (as part of the bargain, he agreed to enroll for a semester at New York University). He rented a room just outside Times Square, set up a small lab in the basement, and began working round the clock on his idea. Years later, Land said, "There's a rule they don't teach you at Harvard Business School: if anything is worth doing, it's worth doing to excess." He persisted, but had no luck with his polarizer idea.

In the face of impossible challenges, where do you go? As we saw in the last chapter, the 42nd Street branch of the New York Public Library. There, Land pored through every book on optics he could find, frequently with a young research assistant he had hired named Helen (Terre) Maislen. Just like Trippe, Land found a clue in the back of an old book.

Sick dogs that were fed quinine to treat parasites showed an unusual type of crystal in their urine. Those microscopic crystals, called herapathite, turned out to be the highest-quality polarizers ever discovered. Scientists had tried for decades, starting in the mid-nineteenth century, to grow the crystals and make useful polarizers out of them. But they failed—the tiny crystals are impossibly fragile—and the field eventually gave up. The discovery had been written out of physics textbooks and the *Encyclopedia Britannica*. Webster's dictionary listed "herapathite" under "obsolete words." The graveyard of unexplained experiments, as Land would soon show, is a great place to find a False Fail.

Land came up with a crazy idea: embed millions of those tiny crystals into some kind of goo (he used a nitrocellulose lacquer) and find a way to get them to line up. After a handful of failures, Land decided to try using a magnetic field to line them up, like a magnet can align small iron filings. He knew of a high-powered magnet at a physics lab at Columbia University. Since he wasn't a student and had no privileges at the university, Land snuck into the building, climbed out onto a sixth-floor ledge, and entered the lab through a window. Land had placed a thin layer of his dark crystal-goo mix inside a plastic cell the size of a quarter. As soon as he placed that cell near the magnet, the dark cell turned transparent. The magnet had done the trick—it aligned the miniature crystals, allowing light to shine through: polarized light. Millions of miniature *Millennium Falcons* streaked toward the plastic cell, but only vertically angled ones could slip through.

It was, he said later, "the most exciting single event in my life." He had created the first man-made polarizer. He was 19 years old.

The following year, Land returned to Harvard. Two months later, he married Terre. He now had access to a lab—but Terre didn't; women at that time were not allowed in labs. So Land would sneak Terre into the physics lab to help him with his experiments. Once again, after a short stint, Land grew restless. Within two years, he abandoned the academic world to start what would soon be known as the Polaroid Corporation.

DISAPPEARING FISH

Land's first big idea was to use his new technology to cut glare from headlights in cars. Headlight glare, at the time, was blamed for thousands of highway fatalities every year. Land realized that coating every car headlight and windshield with a 45-degree filter would allow drivers to see light from their own headlights but not from those of oncoming drivers. To understand why, imagine a child running forward, pretending to be a plane, left arm-wing pointing down to the ground at 45 degrees, right arm-wing pointing to the sky at the same angle. The arms of a second child, running toward the first one, who does the exact same thing, are exactly perpendicular (the four arms form an "X"). The cross-polarized light from an oncoming car can't pass through a driver's windshield for the same reason that a horizontal ship won't pass through a vertical slit. Although Land pleaded with automobile manufacturers for two decades, he could never convince them to adopt his idea.

In the meantime, Land discovered a surprising benefit of polarized lenses. Sunlight reflecting off horizontal surfaces—a still lake or a field of snow, for example—tends to be horizontally polarized. Lenses coated with a vertical-slit film block those reflections far more effectively than ordinary, tinted lenses. The results can be dramatic.

In July 1934, while auto manufacturers were debating and declining his headlight idea, Land arranged a meeting with American Optical, a manufacturer of glasses, at the Copley Hotel in Boston. Land arrived early. A guest would have observed a sharply dressed young man with a piercing stare—one early employee described meeting Land for the first time and feeling that Land "could see into my head. It was really a kind of interesting sensation of having your head briefly searched for content." With his bright eyes, firm jaw, and dark hair slicked and parted, he looked like a movie star. Imagine a young Cary Grant playing the role of an obsessed genius—that's Edwin Land.

Land arrived at the Copley Hotel carrying a goldfish bowl. He asked the desk clerk for a room with western exposure, facing the setting sun. A journalist described what happened next:

> After the bellboy had left, he [Land] placed the bowl on the window sill where it would catch the sun, stood back, inspected it, then moved it so that the reflected glare became more intense. Then he paced nervously and waited for a knock on the door.
>
> As soon as his visitor, an official of the American Optical Company, arrived, he led him to the window and asked him to look into the bowl.
>
> "Do you see any fish?" he said.
>
> The man squinted and shook his head. The reflection from the water was too dazzling.
>
> "Look again," said the young man, holding before the bowl what appeared to be a sheet of smoky cellophane.
>
> The glare was gone as if by magic, and every detail of the idling fish could be clearly seen. The visitor . . . was familiar with every kind of sunglass on the market, but he had never seen anything like this.

Land had his first deal. Sailors, pilots, skiers, and other outdoorsmen soon snapped up the new "polarized" sunglasses, Polaroid's first big hit.

Then the military discovered that eliminating glare from the sun improved gunners' ability to sight aircraft, tanks, and surfaced submarines. The Army and Navy ordered millions of polarized goggles. During World War II, General Patton appeared on the cover of *Newsweek* wearing Polaroid goggles. A *Life* magazine story noted that "every second man in combat" wore them.

The seeds of franchise were growing.

Land soon realized that putting two polarizing filters together produces some striking, and useful, effects. Coat the front of a pair of goggles with a vertically polarizing film; on the back put a polarizer that can rotate inside the frame of the goggles. A tiny handle is attached to that back polarizer, which pokes out of the frame of the goggles at the twelve-o'clock position. When the handle is at twelve o'clock, the two filters line up, and all the light coming in the front goes through the back. But as you rotate the back polarizer, by sliding the handle through ninety degrees toward the three-o'clock position, less and less light makes it through. At exactly ninety degrees—when the front filter is vertical and the back filter is horizontal—no light gets through. Adjustable-shade goggles, which allowed pilots to quickly adjust from low-light to bright-light conditions, were another big Polaroid hit.

Today, if you use a laptop or smartphone or watch something on an LCD screen, you are using a variation of this trick, with a twist, all made possible by Edwin Land's invention.

FROM SMART FISH TO SMARTPHONES

Think of a barn with sliding doors on opposite-facing sides. The back doors slide down from the roof and up from the ground, meeting in the middle, and are closed to a horizontal slit. The front doors slide from the left and from the right and are closed to a vertical slit in the middle. A drone flies through the back opening with its wings horizontal, rotates ninety degrees inside the barn, then flies out through the front opening with its wings vertical.

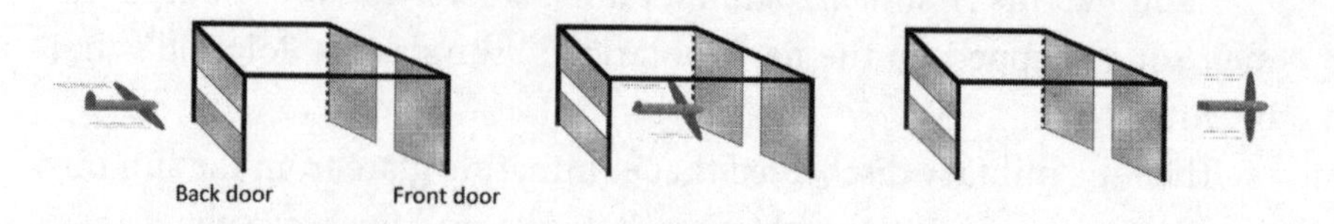

Now suppose the barn came with a switch. Turning on the switch jams any electronics. Drones can't rotate while inside the barn. Any drone flying through the horizontal slit in the back will stay horizontal and crash into the front door. No drone can get through.

LCD pixels work just like those barns.

The back of a pixel on an LCD display screen has a horizontal filter. The front has a vertical filter. Unlike the drone, light cannot rotate on its own traveling through empty space. It needs help. So pixels are filled with a special kind of goo called a liquid crystal, made of billions of microscopic rods, like tiny toothpicks—just like Land's original polarizer. But in this case, the goo is sandwiched between the pixel's horizontal-filter back door and vertical-filter front door. The toothpicks automatically line up horizontally next to the back and vertically next to the front. In between, they form a kind of twisted, quarter-turn spiral staircase, which connects the back and front. The spiral staircase does the work of rotating the light. Light enters through the horizontal opening in the back, travels through the staircase, its polarization rotating by a quarter turn, then flies out the vertical opening in the front and into your eyes. Just like the drone streaking through the barn.

Each pixel, however, comes with a tiny digital switch. Turning the switch on fires up a tiny electric field that scrambles the toothpicks and crashes the spiral staircase. No light can get through. The pixel goes dark. Turning the switch off restores the spiral staircase. The pixel lights up. And there you have it: a digitally controlled on/off light pixel.

The original iPhone screen squeezed in 320 of these digital pixels across and 480 pixels down. Today's smartphone screens and high-definition TVs are made with more than two million pixels.

I mentioned at the start of the chapter that our eyes can't detect polarization. It turns out that many people can train their eyes to pick up one subtle signal. If you look at a white area on an LCD monitor and rotate your head, you may see a small, faint yellow hourglass shape appear and

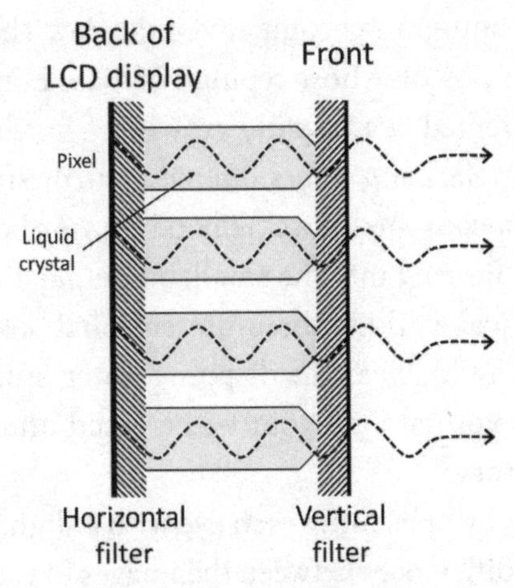

LCDs use polarized light and two filters to create on/off light pixels

then fade. That image, a weird optical effect known as Haidinger's brush, comes from a tiny sensitivity in the back of our eyes to polarized light.

Land's polarizing filters gave rise not only to smart displays and strange tricks, but also to a technology that excited, oddly, both artists and the military. That discovery steered Land toward Polaroid's most famous invention, as well as a 30-year journey that would become the ultimate example of the Moses Trap.

FROM ART FAIR TO WARFARE

In the 1920s and 1930s, Clarence Kennedy, an art history professor at Smith College, an all-women's school in western Massachusetts, produced haunting photographs of sculptures, especially Italian masterpieces. Some described the pictures as more beautiful than the originals. Kennedy cataloged famous collections and advised museums in New York, Boston, and San Francisco. Cities in Italy hired him to restore old monuments (when the Allies began their invasion of Italy in World War II, the US bomber command turned to Kennedy for a list of monuments to avoid). He was a perfectionist, according to a colleague, "but not one of those that irritate."

In the 1930s, Kennedy became obsessed with improving the technology of sculpture photography. Could a two-dimensional image capture the beauty and depth of a three-dimensional form? He spoke with scientists at Eastman Kodak, the dominant photography company of the day. They directed him to a young inventor in Boston whose reputation, based on a new polarizing filter he had just invented, was rapidly growing.

Land quickly realized that his polarizing filters offered a surprising solution to Kennedy's problem, a solution inspired by a childhood toy. As a boy, Land had played with stereoscopes. Peering into the small, binocular-like devices transported you into a magical world of three-dimensional boats and bridges and caves, where you would "hear the dripping water, smell the dampness, fear the darkness as you sat with your legs crossed under you on the chair in the dear old library."

Stereoscopes create those worlds by presenting each eye with a slightly different image. Our brains use the differences between the images in each eye to reconstruct depth: the three-dimensional shape of a sculpture, for example. We perceive ordinary photographs as flat because both of our eyes see exactly the same image. Land realized that to "see" Kennedy's sculpture photography in three dimensions, he would just need to provide each eye with a snapshot taken from a slightly different angle. And he could do that by using his favorite hidden property of light.

First, Land invented a method for fusing two polarized images—one vertically polarized, one horizontally polarized—onto one print. He then made inexpensive glasses with vertical polarizers on one lens, horizontal polarizers on the other. The left eye would see the first image; the right eye, the second. Demonstrating the technique at an optical society meeting held not long afterward in the middle of a presidential campaign, Land projected a fuzzy image on the screen. He asked the audience to put on the special Polaroid glasses, and then he asked Democrats to close their left eye and Republicans to close their right eye. Each group saw their candidate.

Next, Land asked Kennedy for a sculpture to photograph. Land took one picture, moved the camera over a few inches, and then took another. The shift in camera angle captured the difference between what our eyes would see. Land made one image vertically polarized, the other horizontal, and then fused them into one print. When a viewer put on his special polarized glasses, the flat print would burst off the page into glorious three-dimensional form. Land called his new system the vectograph.

In Washington, DC, shortly after his first meeting with FDR, Vannevar Bush heard about Land's vectograph. Within a year, the Army and

Audience for the 3D movie *Bwana Devil* (1952)

Navy were using 3D terrain maps to prepare for battles in Europe. Planes would fly over fields and landing beaches and take pictures a quarter mile apart. With the fused prints, soldiers could see trees or ditches they could use for cover, the contours of hills they would need to climb, and even the fake shadows painted as camouflage on enemy factories.

The technology was likely the first, and possibly the only, example of an art history project weaponized for military use.

Land's 3D still images were soon converted for use in film, which turned into a craze. (At its peak, in 1953, Polaroid was making six million pairs of 3D glasses per week.) Although the novelty of early, low-quality 3D movies wore off, today's 3D films use the same core science Land developed in 1940.

Kennedy's influence on Land and Polaroid continued after 3D photography. He helped grow Land's interest in the art world. Kennedy introduced Land to Ansel Adams, who became a close Polaroid advisor and friend to Land's family, as well as Andy Warhol, Robert Mapplethorpe, Chuck Close, and many others. The art world endorsements added a dash of glamour to the technology, much like Lindbergh and the color spreads of celebrities flying jet planes did for Juan Trippe and Pan Am.

Meroë Morse

Kennedy also contributed one more unusual idea: recruiting Smith College art history majors. Few companies hired women for technical positions in the 1940s and 1950s. Fewer still recruited art history majors and trained them. Kennedy encouraged Land to break both taboos, which became a great advantage for the company; decades before the idea became popular, both Kennedy and Land understood that diversity enhanced creativity. One of Polaroid's most critical technology breakthroughs came from a harpsichord-playing art history graduate from Smith named Meroë Morse, who rose to lead a major research lab for Land. (Morse and Land grew close. A biographer wrote that when Morse died, unmarried, after 20 years at the company working closely with Land, Land "lost a soul mate, a work mate, and a protector. His most severe quarrels with the technical and non-technical sides of his company sprang up after she was gone.")

But Kennedy's singular contribution to the history of business and technology, aside from inspiring Land's interest in 3D images and workplace diversity, was to turn Land's attention to photography.

AN OBVIOUS QUESTION

In December 1943, on a family vacation in Santa Fe, Land went for a stroll with his three-year-old daughter Jennifer. After he snapped some photos of her, she asked him, "Why can't I see them now?" Startled by the question, Land sent Jennifer to her mother. He continued his walk alone, thinking through the problem, turning over the question in his mind, applying insights he had learned from developing 3D photography. Thirty years later,

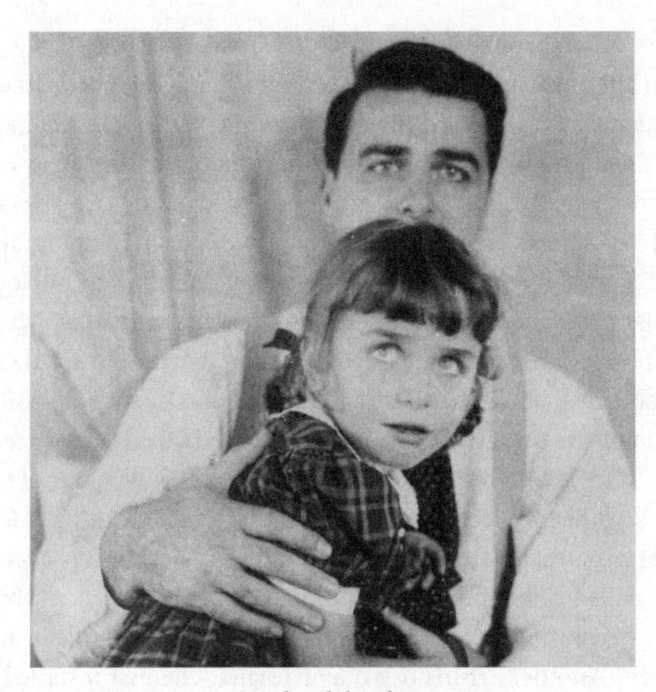

Land and daughter

he recalled the history of his invention to an audience of scientists and engineers: "Strangely, by the end of that walk, the solution to the problem [of instant photography] had been pretty well formulated. I would say that everything had been, except those few details that took from 1943 to 1972."

In traditional film photography, particles of light, called photons, land on film, leaving microscopic residues—a chemical memory. Think of small asteroids striking the surface of the moon, leaving tiny craters. Soaking the film in a developer enhances those residues a billionfold until the familiar negative emerges. It's a negative image because the residues, where the light fell, are dark. To reverse the image and create the usual positive print, you shine light through the film onto white paper; a dark spot becomes white and a white spot becomes dark. Land's insight was to combine those two steps, by developing the negative and the positive at the same time, *inside* the camera, using an ingenious chemical trick.

In Polaroid's instant photography, negative and positive print layers are joined together, like a sandwich, inside the camera, separated by less than one-hundredth of an inch. Attached to the bottom of the sandwich is a small, sealed sac of developing fluid, called a pod. Exiting the camera, the pod passes through a roller, which breaks the sac. Fluid spreads evenly in the thin space between the two layers. The chemistry of that fluid is such that unexposed molecules on the negative, which are light, are suctioned across the thin gap and become dark. The exposed molecules on the negative stay put. Within 60 seconds, the two layers can be pulled apart—presto, an instant print. Jennifer has her photograph.

That "presto," of course, required inventing dozens of technologies and conducting thousands of experiments, the vast majority of which failed—dozens of False Fails and Three Deaths. Land's instructions to take on only those problems that are manifestly important and nearly impossible were his version of "It's not a good drug unless it's been killed three times."

The first to be assigned experiments was Doxie Muller, one of Clarence Kennedy's art history recruits. Land called her every morning at 6:30 a.m. to go over projects for the day. He would review her reports every night. Predawn calls from Land were common. "I had an idea about that problem we've been working on," he would say. "Would you come in and meet me at five?" Another art-history-major-turned-chemist installed a separate phone line in her kitchen: "When the red phone rang, I'd look around to see if my children were killing themselves, and if not, I'd pick it up."

Two years later, in early 1946, results looked promising, but Land felt experiments were moving too slowly. He announced to his team that Polaroid would demonstrate a working camera to the press and industry at the February 21, 1947, Optical Society meeting in New York. His horrified senior team objected; hundreds of technical hurdles remained. Land dismissed the objections. They would present a finished camera in February. The team found another gear.

Land's deadline was about more than injecting urgency into a project team. After Land decided to withdraw from military contract work at the end of the war, sales plummeted from $17 million in 1945 to less than $5 million in 1946, and looked to be less than half that in 1947, threatening the survival of the company. One senior executive recalled, "There was very little income and lots more outgo." Land had bet the company on instant photography.

On February 20, the day before the Optical Society meeting, snow started falling in New York at 4:30 p.m. By the morning it had grown into the largest blizzard in six years. Most of the city had shut down; events all along the East Coast had been canceled. Land and his team waited anx-

ONE-STEP CAMERA IS DEMONSTRATED

Process That Makes Finished Picture in Minute Is Work of Polaroid Company Head

CONVENTIONAL FILM USED

Special Paper Containing Pod of Developer and Hypo Is 'Sandwiched' With Film

A revolutionary new camera, which turns out a finished picture one minute after the shutter is snapped, was demonstrated last night in the Hotel Pennsylvania at the winter meeting of the Optical Society of America by its inventor, Edwin H. Land, president and director of research of the Polaroid Corporation.

The new camera is described in the published program of the Optical Society as "a new kind of photography as revolutionary as the transition from wet plates to daylight-loading film," more than

INVENTOR DEMONSTRATES PHOTO PROCESS

Edwin H. Land showing negative and positive of himself one minute after picture was taken at Hotel Pennsylvania yesterday.

The New York Times

Edwin Land unveils the first instant-print picture

iously to see if the truck from Boston with the camera would make it through in time. It did, barely.

The team quickly assembled the camera for the afternoon presentation. After Land made a brief introduction, he asked the president of the society to come up to the stage. Land aimed, pressed the trigger, peeled apart the layers, and revealed the instant print.

"Everyone went wild," one observer recalled. *Scientific American* described the technology as "one of the greatest advances in the history of photography." The *New York Times* ran a long feature together with an accompanying editorial announcing that all prior photographic inventions were "crude compared with what Mr. Land has done."

That same day, at a special session for the press, Land snapped a self-portrait from the neck up with the new camera. He peeled the print and held it next to his face. At eight by ten inches, it was nearly life size. The *Times* story led with a two-column photo, the inventor staring into the

William Wegman and Andy Warhol Polaroid photography

Life: The SX-70

distance, unsmiling, jaw set. His disembodied head stares sadly out of the page at you, the reader. The haunting image was reprinted endlessly.

* * *

Polaroid sales grew from just under $1.5 million in 1948 to $1.4 *billion* in 1978. For 30 years, Polaroid dominated instant print like Pan Am dominated international travel: by delivering spectacular breakthroughs, year after year, which delighted customers. In both cases, a master P-type innovator at the top fueled those loonshots, which grew the franchise, which, in turn, fueled more loonshots. The wheel in the camera kept on turning. The dangerous virtuous cycle spun faster and faster.

Polaroid followed the first sepia prints in 1947 with black and white (1950); automatic exposure (1960); instant color (1963); non-peel-apart film (1971); the SX-70 all-in-one, foldable camera (1972); sonar auto-focus (1978); and countless other advances in between. For anyone interested in technology, the stories of these inventions are fascinating. To achieve instant color printing, for example, Land and his team invented a new molecule. As a side project, stimulated by a chance observation in the lab, Land invented a new theory of color vision, now called color constancy, which explains why we will see red apples as red even as the color of the light they reflect changes. Land seemed to produce one or two discoveries per year that others would be thrilled to see in a lifetime. One admiring scientist wrote, "Nobel Prizes have been given for less."

With the improving technology came respectability. Instant-print photography, initially considered a toy by serious artists, grew into a new art form. Ansel Adams's 1974 exhibit at the Metropolitan Museum of Art included twenty Polaroid prints. He used Polaroid for both the first presidential photographic commission (Jimmy Carter) and his El Capitan masterpiece. William Wegman's dogs, Andy Warhol's pop, Chuck Close's faces: all were Polaroids.

The technology created not only new art but also new markets. Couples realized their prints would not be seen by technicians at developer labs. And so was born what Polaroid delicately called "intimacy" pictures. Polaroid's growth was helped by a surge in demand for those intimacy pictures, just as years later the internet's rapid growth would be fueled by pornography.

Regardless of the source of the demand, investors rewarded the growing revenues. Wall Street analysts routinely announced Polaroid's shares were overvalued. But the price just kept going up. Fans kept buying and believing.

And then, like Juan Trippe and the 747, the master innovator at the top, creating and anointing loonshots, turned the wheel one too many times.

POLAVISION

In 1888, Thomas Edison wrote, "I am experimenting upon an instrument which does for the Eye what the phonograph does for the Ear." A few years later he used this motion-picture instrument to produce the first American short films. (The shorts included cats in a boxing ring, establishing an enduring principle of human nature: cat videos are always funny.) For the next hundred or so years, movie film was developed more or less like photographic film. A 35-millimeter movie camera captures 24 frames per second onto a film reel negative. The negative is processed in a lab. The biggest difference is that movie film is converted into transparencies through which we project light rather than the familiar solid prints we hold in our hands.

In the mid-1960s, Land began thinking about extending his instant-print technology to film. Instead of one image, over a *thousand* images would need to be processed, instantaneously, without error, for every 60 seconds of film. The process would require reinventing the chemistry of color development and film transparency. A manufacturing plant with entirely new equipment would need to be built on a massive, commercial scale. Several years earlier Land had said, "Do not undertake a program unless the goal is manifestly important and its achievement is nearly impossible." The science and technology behind polarizing filters, instant print, and instant color had all seemed nearly impossible when Land dived in. This new challenge was exactly the kind of P-type loonshot worthy of his mind and energy. And so Land launched what became a ten-year, half-billion-dollar project to create instant-print movies.

* * *

At the 1977 Polaroid shareholders' meeting in Needham, Massachusetts, surrounded by mimes and dancers, in a performance that a *Wall Street Journal* reporter wrote deserved an Academy Award, Land introduced the world to Polavision, announcing "the first public demonstration of a new science, a new art, and a new industry . . . a second revolution in photography." A long-haired dancer in a white sailor suit, with red hat and scarf, emerged on stage and gradually began dancing. Land grabbed a small, elegant movie camera—24 ounces, about the size of a hardcover book—by its angled grip and began to film. After about a minute, he popped out a cassette and inserted it into a rectangular box with a 12-inch screen at one end, called the Polavision player. The player simultaneously rewound and developed the film. Ninety seconds later, the dancer appeared on the screen.

You have to pause to appreciate this, even today in the twenty-first century: processing an entire film negative, thousands of images, inside a consumer tabletop device, without error, while it rewinds, in *90 seconds*.

Technology magazines raved: "The company that seems to specialize in turning impossible concepts into hardware has done it again," wrote *Popular Science*. "The screen suddenly lit up and—to my astonishment—I saw the film I had just made," wrote *Popular Mechanics*. "No Hollywood 'rushes' had ever reached the projection room faster. And no motion picture had ever before been shown without first going to a developing lab." The *Washington Post* wrote, "For the remarkable Land . . . Polavision may well be the highlight of his career."

The new plant produced over 200,000 Polavision machines. The film assembly line began cranking out cassettes. Andy Warhol shot Polavision shorts at his celebrity parties. John Lennon and Yoko Ono made a Polavision home movie with their son Sean. National marketing began in the spring of 1978.

So why haven't you heard of Polavision? Because within a year, the product was dead. Customers weren't buying it. The resolution and quality were superior to magnetic videotape, and the camera, like the SX-70, was a beautiful, lustworthy machine. But customers didn't need that extra resolution for home movies. The elegant design could not overcome the convenience of alternatives. Videotapes and Super 8 film were cheaper, easier, and—in the case of videotape—erasable. With tape, you could record over last week's scene of a cat coughing up a hairball. With instant-print film, you owned a glorious rendering of that scene, in beautiful artistic detail, and could watch it, in real time or surreal slow motion, over and over. But afterward, you had to buy more film, which was expensive. The Polavision camera would set you back close to $2,500 in

2018 dollars, and each three-minute cassette would cost an additional $30. It was too much.

One Wall Street analyst summarized, "This is a product that has much more scientific and aesthetic appeal than commercial significance."

In 1979, Polaroid's accounting firm insisted all unsold Polavision inventory be written off as a loss. It was the public-company equivalent of raising the white flag. Land objected ferociously. The auditors' statement that "the marvelous result of scientific research embedded in Polavision had no utility," Land said, "was accounting jargon, a cruel misuse of language." The board of directors, of course, followed the accountants' recommendation.

Shortly afterward, Polaroid shut down Polavision production permanently. The total project cost for that final year came to just over $200 million. At the urging of the board a few months later, Land resigned as CEO, although he stayed on as head of research. Two uncomfortable years later, he resigned that position as well. He sold his shares and cut all ties to the company he'd founded.

Like Pan Am, a revolving door of subsequent CEOs tried to restore the company's image and edge, to catch up to loonshots nurtured by others. In the case of Pan Am, those were S-type loonshots: new strategies to lower costs or increase revenue per seat. In the case of Polaroid, they were P-type loonshots: video camcorders, home inkjet printers, and, of course, digital photography. As with Pan Am, it was too late.

PHOTONS, ELECTRONS, AND RICHARD NIXON

Traditional photography exploits a chemical reaction. When enough photons (light particles) strike the silver molecules in film, the molecules change form. That creates a *chemical* memory of where photons landed. Under certain special conditions, however, when a photon lands, it can pop an electron out of an atom (the photoelectric effect). The loose electron can be trapped right where the photon landed, like catching a firefly in a jar. Trapped electrons signal their presence with voltage. The voltage forms an *electrical* memory of where photons landed.

In 1969, a small team at Bell Labs created a grid of pixels with just the right conditions to trap electrons popped out of atoms by photons. It was a microscopic grid of jars catching fireflies. They called it a CCD chip. The chips turned out to be up to a hundred times more sensitive than film.

Within a few years, astronomers were using CCD chips to image distant stars. The first commercial cameras using CCDs, for businesses and professionals, appeared in the 1970s. The first consumer digital cameras using CCDs appeared in the mid-1980s.

Polaroid eventually introduced a digital camera in 1996, a decade after similar cameras appeared from Sony, Canon, Nikon, Kodak, Fuji, Casio, and others. It was too late. In 2001, Polaroid filed for bankruptcy.

On the surface—a very public surface—it seems like a brilliant but aging entrepreneur was blindsided by a loonshot: digital photography.

But that's not quite what happened.

* * *

From 2011 to 2015, the US National Reconnaissance Office declassified a trove of documents related to spy satellites. The documents reveal a top-secret, high-stakes drama centered on imaging technology. Well *before* the first astronomers began using CCDs, *before* the first commercial CCD cameras, and *before* Sony and Kodak even began thinking about a consumer market, one person convinced the president of the United States, over the unanimous opposition of his senior military and political advisors, to invest in digital spy satellites. That person was Edwin H. Land.

It was the threat of nuclear war that drew Land into government service. In 1949, the Soviet Union detonated its first atomic bomb. One year later the Cold War turned hot when North Korean forces, backed by the Soviets, invaded South Korea, which was supported by the US. Fears of a nuclear World War III escalated. Shortly after taking office in 1953, President Eisenhower assembled an expert panel, led by the president of MIT, James Killian, to study the possibility of a surprise Soviet attack with nuclear missiles. The panel quickly concluded that the country lacked hard data on Soviet capabilities—missiles, bases, troop movements—and urgently needed new means to acquire that data. The panel needed someone who could not only advise on state-of-the-art imaging science but also anticipate or even design technologies that did not yet exist. Someone who had the strength of personality to challenge generals.

Strength of personality was not a problem for Land. He was quickly selected.

In 1954, Land proposed to Eisenhower the idea of a one-man aircraft with a powerful camera that would fly high and fast. He helped select the camera technology (Itek, Kodak) and the aircraft (Lockheed) for what became the world's first spy plane—the U-2. The plane would play a critical role throughout the Cold War. It was U-2 pictures, for example, that identified Russian missiles in Cuba in 1962.

In 1957, Land and Killian proposed a new idea to Eisenhower. They were concerned about the risks of flying manned aircraft over hostile territory. (The concern was prophetic. In 1960, the Soviets shot down a U-2 plane flying over Russia and captured its pilot.) Land and Killian proposed that instead of flying manned aircraft over enemy land, the country should develop and deploy satellites carrying cameras with giant telescopic lenses pointed at the earth.

Snapping photos in space sounds like a good idea—but how would the photos get back to earth? Land and Killian recommended a system in which the satellites would eject the exposed film in canisters attached to a parachute. Air Force pilots would then fly planes with hooks that fished those canisters out of the sky.

Eisenhower green-lit the program. The president also accepted Land and Killian's recommendation for a new agency, the National Reconnaissance Office, to be jointly managed by the Air Force and CIA.

As the Cold War and Soviet expansion continued, a limitation of the satellite program became increasingly clear. On August 20, 1968, the Soviet Union invaded Czechoslovakia. Satellite film clearly showed a large buildup of Soviet tanks and aircraft on the border before the invasion. But it was old news by the time the Air Force had retrieved the film: the invasion was already over.

Richard Nixon, elected in November, made it clear to his staff that he wanted real-time, not weeks-old, imaging, and he wanted it available by "his second term in office." The scramble rapidly devolved into a bitter battle.

On one side: nearly the entire military leadership and most cabinet members, including the secretary of defense (Melvin Laird), his deputy (David Packard), the head of defense engineering (John Foster), the secretary of the Air Force (Robert Seamans), as well as a future secretary of defense (James Schlesinger) and secretary of state (George Shultz). The military advocated for an incremental solution: add scanners, like fax machines, to existing film satellites. Cameras would take pictures using ordinary film. Those photos would be scanned onboard and transmitted down to stations on earth. To them, the idea of filmless digital photography, using a CCD chip, was too far-fetched, too uncertain. Too much of a loonshot.

On the other side: Edwin Land.

Not surprisingly, the military was winning. By the spring of 1971, a $2 billion program for scanning films inside satellites was gaining steam. At a meeting of the president's intelligence advisory board in April 1971, Land addressed the president directly. He told Nixon that the film-scanner idea was a "cautious step" and that digital technology was a "quantum jump which would give the US an unquestioned technological lead in this field." Land said that bureaucrats "were unwilling to assume large financial risks without strong presidential backing." He explained why digital would work, why the risks were manageable, and why the program was superior to the generals' proposal.

Let's pause: this was the *spring of 1971.* The first papers describing CCDs had been published only months earlier. Sony (which produced the first commercial digital camera), Canon, and Nikon had not even begun to work on digital photography. Land was advocating for digital before *all* of them.

In September, Nixon's national security advisor, Henry Kissinger, informed all parties that the president had decided to proceed with Land's quantum-jump solution. The military's $2 billion program would be terminated. (A National Reconnaissance Office historian attributed Land's eventual success to his "complete understanding of President Nixon's desire to be remembered as a more forceful, incisive, and astute decision maker than his immediate predecessors.")

On December 11, 1976, during what would have been the final days of Nixon's second term (Nixon resigned in August 1974), the Air Force launched the first digital satellite, the KH-11. At 3:15 p.m. on January 21, the day after Jimmy Carter's inauguration, the acting director of the CIA, Hank Knoche, met with Carter and his national security advisor, Zbigniew Brzezinski, in the Map Room of the White House. Knoche spread a handful of black-and-white photos across the table. They were the first live photos taken from space. They showed the president's inauguration ceremony. Better than any words in a briefing document, the pictures explained Land's quantum jump. The US could now look at events around the world "from right up close, virtually as they happened, the way an angel would."

The availability of real-time visual intelligence changed how the US could respond to crises, direct national security operations, and verify arms-control treaties. The much higher sensitivity of CCDs, compared to film, provided images far exceeding anything possible with film-based satellites: they could record the license plate of a truck rather than the outline of a city. By many accounts, the three-hundred-plus digital imaging satellites launched by the NRO have proven to be the most valuable source of intelligence collected by the US over the past 60 years.

Land was not surprised by digital photography. He'd argued for that loonshot in front of the president of the United States. He did so before anyone else was even in the game. In a 1988 ceremony honoring Land, the director of the CIA, William Webster, declared, "The contributions Dr. Land has made to national security are innumerable, and the influence he has had on our present intelligence capabilities is unequaled."

So what happened with Polaroid? Why didn't Land jump on digital for his own company, exploit the head start from his national intelligence connections, and use those advantages to beat Sony, Canon, and Nikon to the punch?

§

FALLING IN LOVE

Moses Trap: When ideas advance only at the pleasure of a holy leader, who acts for love of loonshots rather than strength of strategy

Toward the end of the wild Polavision launch at the Polaroid shareholders' meeting, after Land had finished his presentation and thanked the red-hatted dancer, ushers directed audience members to 20 specially designed "film stations," each equipped with mimes, dancers, and jugglers. Reporters and investors examined the machine, created three-minute instant-print films, and then returned to their seats for the question-and-answer session. Surrounded by dozens of happy performers, Land asked for questions. After some routine comments, an analyst asked, "What about the bottom line?"

Land answered with what became one of the most famous lines of his life: "'The only thing that matters is the bottom line'? What a presumptuous thing to say. The bottom line's in heaven."

* * *

The familiar story of the decline of industry Goliaths begins with decades of success, after which the proud old company grows stale. It loses its hunger. A young upstart, a small David, comes along and slays the lumbering giant with an unexpected weapon. It's a new idea or technology that everyone else overlooked. Some kind of loonshot.

The Goliaths built by Edwin Land, Juan Trippe, and—as we will see in the next chapter—Steve Jobs 1.0 don't fit this picture. Land, Trippe, and Jobs were all master P-type innovators who *never* lost their hunger, their taste for bold, risky projects. Their Goliaths disappeared (or nearly disappeared, in the case of Jobs) because all three followed the same pattern into the same trap.

Each of those visionary leaders created a brilliant loonshot nursery; they achieved Bush-Vail rule #1: phase separation. But they remained judge and jury of new ideas. Unlike Bush and Vail, who saw their role as gardeners tending to the touch and balance between loonshots and franchises, encouraging transfer and exchange, those three master P-type innovators saw themselves as Moses, raising their staffs, anointing the chosen loonshot. In other words, they failed on Bush-Vail rule #2: dynamic equilibrium.

Let's see what we've learned about the Moses Trap and how it seduces even the best of the best.

First: The dangerous, virtuous cycle builds momentum

P-type loonshots feed a growing franchise, which in turn feeds more P-type loonshots. New engines helped Trippe fly farther, faster, with more passengers. Which generated more income, which fed the design of bigger, faster engines. Instant black-and-white prints became instant color prints, which created massive popular demand, which funded the SX-70, with faster pictures, encouraging more photos, which fueled even greater expansion. Faster, better, more.

Second: The franchise blinders harden

Only those P-type loonshots that continue to spin the wheel matter. Trippe saw the new ways of doing business, the S-type loonshots from Bob Crandall and other large carriers, or from local discounters like Pacific Southwest Airlines—but he ignored them. Edwin Land not only saw digital, he dove deep on digital. But he ignored it, for his company, in favor of Polavision. Instant film continued to spin the instant-print wheel. Digital did not.

The P-type loonshot of digital photography took down Polaroid. But there was more to it. That new technology came with hidden S-type loonshots. Land, as we just saw, fully understood the technology. He jumped on its potential and defended the value of digital photography to the highest-ranking generals and political leaders before nearly anyone else in industry had heard of it.

Land and his management team dismissed digital because for 30 years they had made money from selling film: their cameras generated much less income than their instant-print cartridges. With digital, there was no film. "There's no way that can make any money," they said. Land dismissed the new technology because he didn't look for the hidden S-type loonshots: all the ways digital could enable new streams of income. In other words, just like Juan Trippe, he leaned on his strong side—P-type loonshots—and didn't watch his weak side: S-type loonshots.

Third: Moses grows all-powerful and anoints loonshots by decree

Bush and Vail managed the *transfer* rather than the *technology*. They cared for the touch and balance between loonshots and franchises. Land, on the other hand, was the "principal cheerleader and spokesperson" for the Polavision project.

One of Land's admirers, who led various research groups at Polaroid during 20 years at the company, wrote about Land:

> He was boss not only in the corporate sense, but also in the research area, and I suppose that became clearer as time went by. He was not only Chairman and CEO, but also held the title of Director of Research . . . which indicated where his true interest lay. His research decisions would always be governing, never mine.

Not long after the Polavision launch had failed and the product was terminated, Land brought a freelance lighting designer to visit a warehouse full of the instant-movie cameras. The designer asked why Land had brought him to see "this sad landscape."

Land answered: "I wanted you to see what hubris looks like."

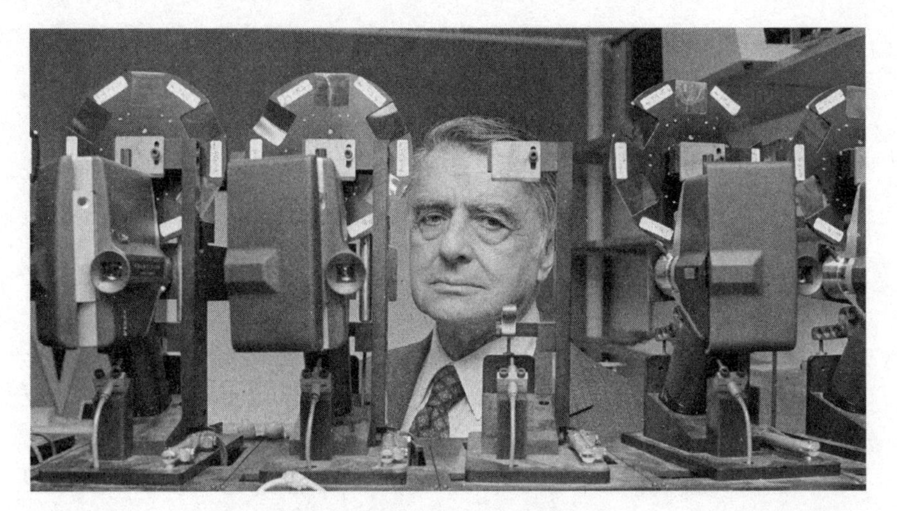

Land and his Polavision machines

* * *

In chapter 1, we used the diagram on the next page to illustrate what Bush and Vail accomplished. They brought aging organizations, with proud franchises rapidly growing stale, to the top-right quadrant. Equally strong research and franchise groups (phase separation) continuously exchanged projects and ideas, with neither side overwhelming the other (dynamic equilibrium).

Land and Trippe succeeded in moving out of the bottom left, but only as far as the bottom right, straight into the Moses Trap.

Land walled off his loonshot nursery from the rest of the company. He banned Bill McCune, the company's head of engineering, as well as anyone else not directly involved in his research, from his private ninth-floor lab. His loonshot nursery produced Nobel-caliber breakthroughs. The franchise group sold millions of cameras. But it was Land who completely controlled which loonshots would emerge, at what time, and under what conditions.

Shifting to the bottom-right quadrant postpones, but does not prevent, the phase transition and decline. A Moses can point to a loonshot and will it to life. But that magic only lasts so long before the wheel stops turning.

The Austro-Germanic school of fatalism (Spengler, Schumpeter) says that decline is inevitable. Empires will always ossify, a David will always

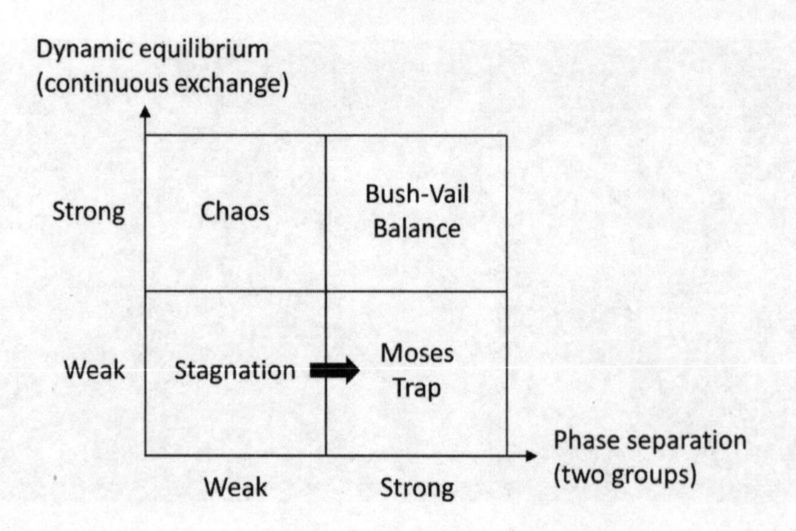

rise to slay Goliath, and so it goes. Is that cycle of creative destruction truly inevitable? What's an empire to do?

Bush and Vail understood that the doomsday cycle is *not* inevitable, and that the best chance for sustainable, renewable creativity and growth comes from bringing an organization to the top-right quadrant: separate phases connected by a balanced, dynamic equilibrium.

But how do we get there?

5

Escaping the Moses Trap

Buzz and Woody rescue a 747, invent the iPhone, and explain system mindset

"Steven P. Jobs is back," declared the *New York Times* on October 13, 1988, describing the first product launch for Jobs's new company, NeXT Inc. Three years earlier Jobs had parted ways in an ugly divorce with the company he had cofounded: Apple Computer.

Three thousand people gathered at Davies Symphony Hall in San Francisco to see Jobs unveil NeXT's computer.

"I think together we're going to experience one of those times that occurs only once or twice a decade in computing," Jobs announces, opening the event. Jobs is wearing a boxy dark suit, skinny tie, and shaggy hair—a hyper-caffeinated fifth Beatle. "This is a revolution," he says.

The article continues:

> Mr. Jobs is known for his dramatic product introductions and he and his company took advantage of intense interest in the computer community about both him and his new machine.
>
> He stood alone on a dark stage with just the computer and a vase of flowers, a huge screen behind him, and took the new machine through its paces. He demonstrated how it could record and send voice messages, play music with the quality of a compact disk and instantly retrieve quotations from the complete works of Shakespeare stored on its optical disk.

Wrapping up the two-hour demonstration of processors and ports and object-oriented programming, Jobs brings his long fingers together into Zen greeting position and pauses.

"One of my heroes has always been Dr. Edwin Land, the founder of Polaroid," Jobs declares. "He said that he wanted Polaroid to stand at the intersection of art and science. We feel the same thing about NeXT. And of all the things that we've experienced together here today I think the one that strikes closest to the soul is the music."

With that, Jobs introduces Dan Kobialka, a principal violinist of the San Francisco Symphony. Kobialka approaches the NeXT computer, playfully taps it with his bow, and begins a thundering five-minute duet. Machine joins man for Bach's Violin Concerto in A Minor. When Kobialka finishes and looks up, a third spotlight rises on Jobs, holding a red rose. The crowd erupts in a standing ovation.

Replace the violinist with a red-hatted dancer, and it's the Polavision launch.

EIGHT MEGABYTES OF SEXUAL SATISFACTION

The popular press gushed. A *Newsweek* cover declared that Jobs "put the 'wow' back in computers." The *Chicago Tribune* noted that the launch event was "to product demonstrations what Vatican II was to church meetings." Another headline simply read: "Eight megabytes of sexual satisfaction!" The launch also inspired competitive trash-talking. When asked if Microsoft would create software for the new machine, Bill Gates answered: "Develop for it? I'll piss on it." He dismissed the technology ("anybody can write Sony a check") and the sleek, all-black design ("if you want black, I'll get you a can of paint").

Five months after the launch, NeXT announced a partnership with the largest computer retailer in the country, Businessland. David Norman, the president of the retailer, projected $150 million in sales in the first twelve months, an unprecedented figure. At a gathering of top Businessland staff, Jobs pumped them up to "kick the shit out of some people!" One attendee described a scene with Businessland sales managers soon after: "Picture grown, smart adults, standing on their chairs, screaming, they were so excited."

To build the machines, Jobs insisted on a state-of-the-art, fully automated factory with art gallery walls and lighting, designer bathroom fixtures, and high-end leather furniture. One journalist described the factory as ready for a cover of *Architectural Digest*.

IBM and Apple sold millions of personal computers a year. Sun sold over 100,000 workstations a year. Jobs had designed his factory for billions of dollars in sales. Over the course of one year, Businessland sold fewer than four hundred NeXT machines.

Like Polavision or the Boeing 747, the NeXT Cube was a beautiful, technologically remarkable, wildly expensive machine—with no customers. The new optical drives had many times the memory of magnetic drives or floppy disks. But competitors offered more convenience, more useful applications, and lower costs. The summary of Polavision—"a product that has much more scientific and aesthetic appeal than commercial significance"—applied equally well to the NeXT computer.

"We saw some new technology and we made a decision to risk our company," Jobs had announced at the launch event, speaking of the optical drives. Scott McNealy was the CEO at Sun, one of NeXT's chief competitors. McNealy recognized that $10,000 machines were not impulse buys influenced by glitzy marketing events and sleek designs. The large customers that could afford them wanted practical machines, with swappable parts, using reliable hardware.

Jobs spoke of love of loonshots. McNealy acted on strength of strategy. Sun grew to over $3 billion in sales. Two years after the launch, NeXT's retail partner, Businessland, went out of business. Its big bet on NeXT was not the only body blow, but it contributed.

By April 1991 two of Jobs's cofounders at NeXT had resigned. In June, Ross Perot, NeXT's largest individual investor, resigned from its board of directors, stating, "I shouldn't have let you guys have all that money. Biggest mistake I made." Over the next few months, the company borrowed from banks to make payroll. With NeXT on the edge of bankruptcy, Jobs went to its partner and largest investor, the Japanese company Canon, which manufactured both the computer's optical drive and its printer. Canon wrote a check, and did so again two more times over the next year, before finally drawing a line. By early 1993, nearly all the vice presidents at the company, including all five of Jobs's original cofounders, had left.

A *Forbes* article stated, "There are very few miracle workers in the business world, and it is now clear that Steve Jobs is not one of them."

WHEN MOSES DOUBLES DOWN

The facts of Jobs's forced exit from Apple in 1985, and his path to the mess at NeXT, have been well laid out. In 1975, Steve Wozniak combined a microprocessor, keyboard, and screen into one of the earliest personal computers. Jobs convinced Wozniak to quit his job and start a company. After some initial success with their Apple I and II, however, competitors quickly passed Apple by. In 1980, Atari and Radio Shack (TRS-80) sold roughly seven times as many computers as Apple. By 1983, Commodore dominated the market, with the IBM PC, launched only two years earlier, a close second. Apple's share had dropped to less than 10 percent and was shrinking rapidly.

Apple's attempts to win back the spotlight with the Apple III and the Lisa, projects led by Jobs until he lost interest (in one case) or was kicked off (in the other), flopped. The legendary Super Bowl ad in early 1984 for a new Apple product, called the Macintosh, created tremendous publicity and an initial burst of sales. But the computer was painfully slow, had no hard drive, and frequently overheated (Jobs had insisted on no fan, to keep it quiet). In a year in which IBM and Commodore each sold over two million computers, Macintosh sales dwindled to less than ten thousand per month.

Even more dangerous to the company's future than its string of failures, however, was its string of exits.

A stream of departing employees signals serious dysfunction. As mentioned earlier, after founding what became Bell Labs, Theodore Vail said that no group "can be either ignored or favored at the expense of the others without unbalancing the whole." Vannevar Bush, during the Second World War, took every chance he could to emphasize his respect for the military, even as he spent nearly all his time with scientists like himself. Loving your loonshot and franchise groups equally, however, requires overcoming natural preferences. Artists tend to favor artists. Soldiers tend to favor soldiers.

Jobs proudly and publicly referred to his team, working on the Macintosh, as artists. He referred to the rest of the company, developing the Apple II franchise, as bozos. Apple II engineers took to wearing buttons with a circle and line running through an image of Bozo the Clown. Wozniak, an engineer with the demeanor of a teddy bear, was widely beloved at the company and in the industry. He resigned, openly complaining about the demoralizing attacks. Departures in the Apple II group became so common that one joke ran, "If your boss calls, be sure to get his name." The toxicity spread. Key designers on the Macintosh side soon began leaving as well.

It didn't take long for the Apple Board of Directors and its recently hired CEO, John Sculley, to conclude the dysfunction was not sustainable. Jobs was stripped of operating responsibility in the spring. Jobs discussed with them the idea of staying and creating a small unit to develop new technologies he'd heard about. Touchscreens. Flat-panel displays. A superpowerful graphics computer from a group of quirky engineers in Marin County, just north of San Francisco. In the end, however, Jobs decided to go. He resigned officially, to start NeXT, in September 1985.

The idea of a superpowerful graphics computer, however, stayed with him.

After Jobs left Apple, the remaining team, led by John Sculley, fixed the most glaring Macintosh flaws. They restored the fan, added a hard drive, and increased the memory (which improved speed). Sales turned around, and the product became a hit. Jobs was soon hailed, retroactively, as a master product innovator. He had created the Apple II and the Macintosh. He had brought personal computing, the graphical user interface, and the mouse to the masses. *Playboy* and *Rolling Stone* interviewed him. He made the covers of *Time, Newsweek,* and *Fortune.* One business magazine, *Inc.*, named him Entrepreneur of the Decade.

As NeXT began to struggle, even as Jobs's star was rising, several employees at NeXT, as well as executives from Compaq and Dell, approached Jobs with an idea: get out of hardware. NeXT's software was excellent. Its graphical interface and programming tools were more elegant and powerful than Microsoft's DOS and early Windows. Jobs could offer PC makers an alternative to Microsoft, which they desperately wanted. In return, the PC makers could offer NeXT something it desperately needed: a future.

The idea of switching from hardware to software was a classic S-type loonshot. Jobs had risen to fame selling hardware. Bigger, faster, more, every year. The stars of the day—IBM, DEC, Compaq, Dell—sold shiny machines stamped with their famous logos. Everyone knew there was no money to be made in software; the money was in hardware.

And dozens of stories hailed Jobs as the master P-type innovator of his generation. Just like Edwin Land and Juan Trippe before him.

Abandon hardware? Not this Moses.

In fact, Jobs had already doubled down. Not long after he left Apple, Jobs got back in touch with the team of engineers in Marin County developing a graphics computer. Why bet on just one bigger, faster machine if you could have two? He bought their business and left them alone to build an even more powerful computer than NeXT.

Jobs had no idea that those engineers held the key to rescuing him from the Moses Trap. And it would have nothing to do with their machine.

ISAAC NEWTON VS. STEVE JOBS: A BRIEF INTERLUDE

Stories of great breakthroughs tend to coalesce around one person, one genius, and often one moment. Those stories are fun to tell and easy to digest. Occasionally they are true. More often, they contain a kernel of truth, but omit a much richer and more interesting picture.

Isaac Newton, for example, is often celebrated for discovering universal gravity, explaining the motion of the planets, and inventing calculus. But well before Newton's *Principia*, it was Johannes Kepler who first suggested the idea of a force from the sun driving the motion of the planets, Robert Hooke who first suggested a principle of universal gravity, Christiaan Huygens who showed that circular motion generates a centrifugal force, many who used Huygens's law to derive the now-familiar form of gravity, Giovanni Borelli who explained the elliptical motion of Jupiter's moons using gravitational forces, John Wallis and others who created the differential mathematics Newton used, and Gottfried Leibniz who invented calculus in the form we use today. That story is harder to tell than the apple falling on Newton's head.

Hooke suggested to Newton how gravity can explain planetary motion. Hooke's suggestions launched Newton on the path to his masterpiece, *Principia*. Although Hooke suggested some of the initial ideas, he did not have the skills to create a complete system. Newton did. Newton was a great synthesizer, just as Jobs was a great synthesizer.

Isaac Newton had Robert Hooke. Steve Jobs had Jef Raskin. Robert Hooke, in his spare time, designed bat-like flying wings, developed sprung shoes to bounce around London in twelve-foot-high leaps, and investigated the uses of marijuana ("the Patient understands not, nor remembereth any Thing that he seeth . . . yet is he very merry"). Jef Raskin, in his spare time, designed and built remote-controlled plane kits, taught harpsichord, conducted an opera company, and filed patents on packaging design. Like Hooke, Raskin was a bit of a dabbler.

In 1967, Raskin, then a 24-year-old engineer, submitted a PhD thesis arguing that computers should have graphical interfaces and that their usability was more important than their efficiency. Both were radical ideas at the time, when monolithic mainframe computers dominated. In the early 1970s, Raskin ended up as a visiting researcher at Stanford and Xerox PARC. At PARC, he saw scientists create the first graphics-enabled personal computer, the Alto, with a bitmapped screen, a graphical interface, icons, and a mouse. (PARC failed to commercialize any of those technologies. For more on PARC as an example of how *not* to escape the Moses Trap, see the summary section at the end of this chapter.)

Raskin joined Apple in 1978, one year after Jobs and Wozniak started the company. Not long afterward, he launched a project to create an easy-to-use, inexpensive, graphics-enabled, small-footprint computer based on the Alto. He called it the Macintosh project. Jobs and others at Apple tried to terminate the project, so Raskin encouraged them to visit Xerox PARC and see for themselves. They did and were converted. Eventually Jobs shoved Raskin aside and took over the project.

Raskin launched the original Macintosh project and suggested some of its core ideas to Jobs. But he did not have the skills to develop those ideas into a complete system. Jobs did. Jobs was a great synthesizer.

Newton and Jobs also treated their precursors in a similar fashion. Newton tried to crush Hooke and bury his contributions (including, allegedly, losing the only known portrait of him). Newton described Hooke, in language that stuck for three centuries, as "a man of strange, unsociable temper." Jobs described Raskin as "a shithead who sucks."

In an interview after Jobs's death, Bill Gates said, "Steve and I will always get more credit than we deserve, because otherwise the story's too complicated." He added, "But the difference between him and the next thousand isn't like, you know, God was born and he came down from the hill with the tablet." I believe Gates may be mixing Jesus and Moses metaphors. But his point was clear.

The richer stories do more than just correct cartoonish summaries—Newton discovered gravity; Jobs created the Mac—or humanize deities. The richer stories help us understand how the forces of genius and

"And he came down from the hill with the tablet"

serendipity come together to produce great breakthroughs. The true histories, rather than the revisionist histories, contain the clues from which we learn how to make the forces of genius and serendipity work for us rather than against us.

The first of those clues, in the case of Steve Jobs, appeared 36 minutes into a 1976 film starring Peter Fonda and Blythe Danner.

FUTUREWORLD

Scene: Spaceship cockpit. Décor: 1970s. Lots of computers with blinking lights. Scientist 1, in white lab coat, strides into frame. Monotone computer voiceover: "Hyaline and synovial readouts recorded."

Scientist 1: Status?

Scientist 2, seated, examining monitor: We're completing the gross body series. We'll start molecular studies in one hour.

```
1: All right. Did you alter their food?

2: Yes, sir, we should have 4 to 6 hours.

1: I want all thermal x-ray and electrochemical studies
finished by tonight.

2: That's not much time.

1: It'll have to do. Our Mr. Browning is getting much
too curious.

2: I have a holograph in my screen. Restructuring.
```

A translucent white three-dimensional image of a left hand appears, fingers extended upward, and slowly rotates. The three leftmost fingers curl down. Then the wrist curls down, the thumb tucks in, and the hand rotates until the forefinger points directly out of the screen, at you, the viewer.

In 2011, the Library of Congress selected this clip as one of 25 to be added to the National Film Registry. Not the film itself, *Futureworld*—which somehow brought together, in one movie, sex robots, scenes of medieval jousting, and Yul Brynner dressed like a gay cowboy—but the three-dimensional hand. The rotating hand was the first 3D computer-generated image to appear in film. It was made by a physics major turned computer graphics programmer at the University of Utah named Ed Catmull.

Academic disciplines tend to flower on different campuses at different times, like flash mobs. In the 1970s, a mob of young computer graphics pioneers flashed on the campus of the University of Utah: Jim Clark, who would go on to create Silicon Graphics; Nolan Bushnell, who would start Atari; John Warnock, who would create Adobe; and Alan Kay, who would help create the first graphics-enabled personal computer, the Alto, at Xerox. Joining them was Catmull, a mild-mannered Mormon graduate student who would cofound the greatest animated film company of his time.

At Utah, Catmull created the 3D hand for a class project. He and his thesis advisor, a graphics pioneer named Ivan Sutherland, took it to Disney. Walt Disney had been a boyhood idol for Catmull, who had dreamed of becoming a Disney animator. Catmull approached the animation building like visiting a shrine. Disney, however, passed on his technology. And there would be no animation job offer for Catmull.

Over the next decade, Disney, an empire built on animation, would dismiss a remarkable string of animation technologies invented by the Utah graphics alumni; just as Xerox, an empire built on office productivity, would dismiss a remarkable stream of loonshots that transformed office productivity, invented by its subsidiary Xerox PARC.

Meanwhile, Catmull had finished his PhD and needed a job. He had invented an important mathematical tool for mapping images and textures onto objects: it could project a picture of Mickey Mouse, say, onto the surface of a tennis ball. He had created the first 3D animated image to appear in film. But no one, it seems, was interested. Catmull was 29, married, and had a two-year-old son. He ended up in Boston at a computer software company.

Until a man called about a tuba.

FROM TUBBY TO PIC

In the 1960s, Alex Schure, a fast-talking, eccentric millionaire, acquired a handful of mansions near the northern coast of Long Island, New York, and turned them into the campus of a private trade school he called the New York Institute of Technology. The school was intended, initially, for people who couldn't enroll anywhere else. Since many of his students needed remedial help in math, Schure hired a comic-book artist to draw their math lessons. That went well so he hired animators to convert those cartoons into a film. The film won a gold medal in the New York International TV film festival. As eccentric millionaires with one success are inclined to do, Schure concluded he was an expert, a proven filmmaker. He would write, direct, and produce his next project. He called it *Tubby the Tuba.*

Schure hired a hundred animators to begin work on *Tubby*, but he soon realized that drawing each frame by hand was a tedious, painstaking process. A search for better technologies for *Tubby* led him to Utah, which led him to Catmull, and a phone call. Would Catmull accept a large amount of money to set up an independent research lab, hire a team, buy whatever equipment they needed, no strings attached, just develop great animation technology? Catmull quit his job and joined Schure in Long Island.

One of Catmull's first hires was Alvy Smith, a big, long-haired Texan with a PhD in computer science. Smith had taught at New York University for five years, then decided to leave academia and move to Berkeley, California, with no plans. He eventually found his way to Xerox PARC, where he worked on color displays and graphics software (the first computer painting tool was also developed at PARC). Less than a year later, however, Xerox terminated the project and let him go. His supervisor explained, "Color is not a part of the office of the future."

By then, Smith was hooked on the potential of computer graphics. He was desperate to find a way back in. He soon heard about "a madman in Long Island" building a lab. Smith spent the last money he had on a plane ticket, visited NYIT, and was immediately hired. The Utah Mormon and the Texas hippie settled into a garage—the converted two-story, four-car garage of the former Vanderbilt-Whitney estate—and began building the most advanced computer graphics lab in the country. It marked the beginning of a computer graphics dynasty, "a marriage of the house of Xerox and the house of Utah," Smith wrote.

In the spring of 1977 at a private theater in Manhattan, Alex Schure proudly unveiled his finished film to his team. At the end of the screening, one of the film's animators quietly said, "Oh God, I've wasted two years of my life." Catmull described the film as a train wreck. The production was amateurish. Catmull and Smith saw that Schure had no instinct for story

Tubby the Tuba

or character. They recognized that Schure would be no Walt Disney, and that computer-generated film to rival live action, their dream, would never emerge from the Vanderbilt-Whitney garage.

Fortunately, not long afterward, another mogul looking for better technologies for his movies called. His film had premiered one week after *Tubby*, and he was already working on a highly anticipated sequel. But drawing light sabers by hand, frame by frame, duel by duel, was taking too damn long.

A giant swooshing sound could be heard along the coast of Long Island as the graphics group abandoned the maker of *Tubby* for the idol of geeks everywhere, the maker of *Star Wars*. The team relocated to a nondescript office building in Marin County, California, home to George Lucas's film production operation. Over the next five years, the Lucasfilm Computer Division, as they were soon known, originated much of the software and hardware that has transformed filmmaking over the past forty years: 3D rendering, digital editing, optical scanning, laser film printing, and, of course, astonishingly realistic computer-generated imagery, CGI. As a teenager, I saw the first scene they made that was used in a feature film: the genesis effect in *Star Trek II: The Wrath of Khan* (1982). It almost made up for the tears I shed for Spock.

The powerful graphics computer built by the Lucasfilm group to create these effects needed a name. Smith suggested "Pixer," for pixel + laser. A colleague in the graphics group suggested something more high-tech, like radar, or astronomical, like quasar or pulsar. They converged on the Pixar Image Computer, which was soon called simply the PIC.

* * *

In 1985, while Steve Jobs was in the middle of his prolonged unhappy exit from Apple, a colleague, Alan Kay, suggested that he look at the PIC. Kay had been one of the early personal computer pioneers at Xerox PARC before joining Apple. Kay had overlapped with Catmull at Utah and Smith at Xerox. He had heard from them that Lucas, who had recently divorced and needed cash, was looking to sell the group.

Jobs had been dreaming up plans for NeXT, but suddenly there was PIC. It was big, fast, powerful, and enormously sexy (the group worked with *both* George Lucas and Steven Spielberg). It was also very expensive: a $100,000 machine. In the fall of 1985, in an interview with the

Wall Street Journal, Lucas explained the many potential uses for the PIC: imaging in radiology, oil and gas exploration, automobile design, and so on. "The movie business turns out to be a minuscule market compared to others we now find ourselves involved with," Lucas said.

"It's like we designed this very sophisticated race car capable of doing all sorts of amazing and complex feats on the race track and then come to find out that a huge segment of the population wants to use it to commute to work." It was a good story, from a legendary storyteller, looking to unload a business.

Jobs was convinced. First, he tried to interest Apple board members in acquiring PIC for the company. They rejected the idea. Later that summer, as his relationship with Apple disintegrated, Jobs proposed to Catmull and Smith that he acquire their operation and run it. Listening to Jobs, Catmull recalled, it became clear that "his goal was to build the next generation of home computers to compete with Apple." They had no interest in that fight, and declined. Catmull, who had recently separated from his wife, picked up on the bitterness. He told Smith, "We don't want to be the first woman after the divorce."

Toward the end of 1985, after nearly two dozen firms (including Disney) had passed, the president of Lucas's studio, Doug Norby, decided he would shut down the computer group at the end of the year if they couldn't find a buyer. Fortunately, in November, Catmull and Smith convinced the Dutch electronics company Philips, which wanted the medical imaging applications, and the auto manufacturer General Motors, which wanted the computer-aided design business, to jointly purchase the graphics group. One week before signing, however, the deal fell through. Ross Perot, the head of the computer division at GM, had spearheaded the drive to acquire the graphics computer. Right around the time of the deal, GM announced the acquisition of Hughes Aircraft for $5.2 billion. Perot was furious and insulted the board, both privately and publicly: "How could GM justify spending billions on a communications satellite operation when it couldn't even build a reliable car?" In return, the GM board members withdrew their support for his computer deal (the following year they got rid of Perot).

Jobs heard about the buyout falling through. He called Norby and said he was still interested. Jobs needed to convince not only Lucas's studio chief to sell the group to him at a reasonable price, but also Catmull and Smith to continue the project, working for him. By then, Jobs had started NeXT. He told Catmull and Smith that they could run their own show. They could stay in Marin County, a few hours north of Jobs and NeXT. Catmull would be the CEO. Catmull and Smith, who were out of options at that point, agreed. Norby accepted Jobs's lowball offer to buy the whole unit.

And so Jobs became the principal investor, and largest shareholder, of the Lucasfilm Computer Division, which was renamed Pixar, Inc.

"Look What Steve Jobs Found at the Movies!" read the *BusinessWeek* headline.

FIRE-HYDRANT YEARS

Jobs bought the group for its big computer. "Image computing will explode during the next few years, just as supercomputing has become a commercial reality," he said in announcing the purchase. "This whole thing has the same flavor as the PC industry in 1978." Jobs terminated Pixar's one ongoing film project, directed the group to open PIC sales offices in seven cities, and added hardware sales staff, growing the company from 40 employees to 140.

After two years, fewer than two hundred machines had been sold. The promise of PIC turned out to be more fantasy than fact. Much of the CGI work could be accomplished using Pixar software on less expensive and more versatile workstations, like those made by Sun or Silicon Graphics. The PIC hardware wasn't needed. In 1986, to highlight the potential of computer-generated animation, Pixar created their famous *Luxo Jr.* clip. Disney's head of animation said that "Luxo the lamp had more emotion and humor in a five-minute short than most two-hour movies." The clip, now part of the Pixar logo, was made using workstations rather than the PIC.

Like NeXT, like Polavision, like the Boeing 747, the PIC was a beautiful, turbo-powered, wildly expensive machine—with no customers. Once again, love of loonshots had triumphed over strength of strategy, just as it had with Juan Trippe and Edwin Land. Only Jobs, unlike the other two, had doubled down on the Moses Trap.

After two more years and over $50 million invested, Jobs finally pulled the plug on the PIC. In April 1990, Pixar sold its hardware business to a California-based technology company, Vicom Systems. Vicom went out of business soon afterward. Jobs shrank the company back down to 40 or so employees, laying off all the people he had insisted the company hire.

Pixar was crumbling; NeXT was floundering, and Jobs was finally running low on cash. Jobs tried to shut down the animation group at Pixar—only five employees—but Catmull and his team resisted. Jobs tried to sell Pixar, but he couldn't find a buyer at acceptable terms. Later he described that time as being in "ankle-deep shit." He stayed home rather than go to work.

Years ago, when I was feeling down about bad news my company and I had just released, an advisor, who had retired from several decades of running a large public company, put his arm around my shoulder and said, "Some days you're the dog. Some days you're the fire hydrant." For Jobs, these were fire-hydrant years.

In the world of biotech, struggling startups often try to buy time by selling tools and services to their much larger, richer cousins, the big pharmas. The goal is to survive long enough for the internal team to create a product—a strikingly original drug candidate.

And that's exactly what Pixar did in the world of film. It sold tools and services to a much larger, richer cousin—Disney—and survived long enough for the internal team to create a strikingly original product.

With Pixar, however, Jobs gained not only time—and a product he never expected—but an idea. He found a different way to nurture loonshots.

BUZZ AND WOODY SAVE THE DAY

On the evening before Thanksgiving 1995, at the El Capitan Theatre in Los Angeles, the lights dimmed and curtains rose on an animated toy astronaut named Buzz Lightyear and a toy cowboy named Woody. Pixar's *Toy Story*, the industry's first fully computer-generated feature film, became the highest-grossing film in the country for three weeks in a row. The film still has a 100 percent rating on Rotten Tomatoes. Reviews at the time described it as "visually astounding," "the rebirth of an art form," "the dawn of a new era." Conceived and directed by John Lasseter, the same artist behind the *Luxo Jr.* clip a decade earlier, the film began Catmull and Lasseter's reign as the greatest animators since Walt Disney and made Jobs, who had previously tried to get rid of the animation unit, suddenly *very* interested in this new art form. The success came with another side effect: it made Jobs a billionaire.

The film was the culmination of a ten-year relationship with Disney. While still at Lucasfilm, Catmull and Smith had convinced Disney to purchase a handful of PICs to automate animation. Disney saw Pixar's short clips win standing ovations at graphics conventions and concluded the team could make a feature film. In 1991, after failing to hire Lasseter away from the group, Disney signed a three-picture deal with Pixar. *Toy Story* was the first of the three.

In the months leading up to the premiere, Jobs worked with bankers to prepare Pixar for an initial public offering of stock. IPO preparation consists, among other things, of drafting a prospectus, a document distributed to investors that describes the company. The cover of Pixar's prospectus featured a smiling Buzz Lightyear leaping out of a computer monitor. I've drafted many prospectuses and participated in many offerings. None had a giant image of a lovable action toy on the cover. None was as perfectly timed.

Pixar's offering, one week after the film premiere, exploded into an investor frenzy. The stock began trading at 250 percent above the bankers' initial estimates. By end of day the company was valued at $1.5 billion. Jobs owned 80 percent. His stake was worth $1.2 billion. Not long before, his ability to continue to support any of his ventures had been in serious doubt.

Earlier I mentioned that Newton and Jobs were great synthesizers. Newton brought together planetary astronomy, laws of motion, differential mathematics—ideas developed by others—and synthesized them into a coherent whole the world hadn't seen. Jobs brought together design, marketing, and technology into a coherent whole, as few others could do. But he was missing a key ingredient. Like Land before him, who brought similar skills together, Jobs had led only as a Moses.

Which is why the most valuable gift that Jobs received—from the perspective of Apple product lovers today—was not the financial reward of his Pixar investment. It was seeing the Bush-Vail rules in action. He learned a different model for leading, for how to nurture loonshots and grow franchises while balancing the tensions between the two.

That missing ingredient became the key to his third act, when he returned to hardware and revived his previous company—along with the entire American consumer electronics industry.

MOVIES AND DRUGS: A BRIEF INTERLUDE

Pixar's story has a great plot: a small, struggling company, dismissed by nearly all the major players in the industry, is saved by a partnership. The partnership produces an industry-transforming hit. The hit launches a wildly successful public offering. The offering finances a staggering run of new hits: *Monsters, Inc., Finding Nemo, The Incredibles, Cars, Ratatouille, Wall-E, Up, Inside Out*, and others.

The Pixar story is a marvelous remake. Fifteen years earlier, in 1978, a tiny, profitless company called Genentech, developing an unproven new technology called genetic engineering, which was dismissed by nearly all the incumbent players in the industry, signed a partnership with a large pharma company. Pixar's technology automated a manual process and allowed animators to create a new kind of film. Genentech's technology automated a manual process and allowed scientists to create a new kind of drug.

Genentech's public offering was perfectly timed and beautifully marketed, just like Pixar's. The wildly oversubscribed offering closed on October 14, 1980. The stock began trading at 200 percent above bankers' initial estimates. Pixar's IPO marked the birth of a new art form. Genentech's IPO marked the birth of a new industry—the biotechnology industry. The successful offering financed a staggering run of hits: Herceptin (for breast cancer), Avastin (for colon, lung, and brain cancers), Rituxan (for blood cancers).

Both Genentech and Pixar—like any good drug-discovery company or film studio—learned how to balance both loonshots and franchises because they *had* to. There are no other kinds of products in movies and drugs.

In the biotech world, probably no company did it better than Genentech. In 2009, when it was sold to Roche, the company was valued at just over $100 billion. In the film world, probably no studio did it better than Pixar. From 1995 through 2016, Pixar released 17 feature-length films. The films *averaged* over half a billion dollars in gross sales each. Their median Rotten Tomatoes score is an astounding 96 percent.

BALANCING UGLY BABIES AND THE BEAST

Ed Catmull, from Pixar, refers to early-stage ideas for films—loonshots—as "Ugly Babies." The language is new, but the idea goes back centuries. In 1597, the philosopher Sir Francis Bacon wrote, "As the births of living creatures are at first ill-shapen, so are all Innovations, which are the births of time." Here is Catmull describing the need to maintain the balance between loonshots and franchises—"the Beast"—in film:

> Originality is fragile. And, in its first moments, it's often far from pretty. This is why I call early mock-ups of our films "Ugly Babies." They are not beautiful, miniature versions of the adults they will grow up to be. They are truly ugly: awkward and unformed, vulnerable and incomplete. They need nurturing—in the form of time and patience—in order to grow. What this means is that they have a hard time coexisting with the Beast. . . .
>
> When I talk about the Beast and the Baby, it can seem very black and white—that the Beast is all bad and the Baby is all good. The truth is, reality lies somewhere in between. The Beast is a glutton but also a valuable motivator. The Baby is so pure and unsullied, so full of potential, but it's also needy and unpredictable and can keep you up at night. The key is for your Beast and your Babies to coexist peacefully, and that requires that you keep various forces in balance.

Keeping the forces in balance is so difficult because loonshots and franchises follow such different paths. Surviving those journeys requires passionate, intensely committed people—with very different skills and values. Artists and soldiers.

The many rejections of the first James Bond movie, *Dr. No*, for example, are typical for original films, just as the convoluted history of the first statin is typical for breakthrough drugs. Bond was too British for American studios; Fleming's novels were "not even good enough for television." Fleming gave up trying to sell the film rights to studios after a decade or so—nine novels into his series—and granted the rights to a pair of dubious producers. One had just bankrupted his production company. The other had limited film experience. The partnership did not start well. The first writers changed the villain of *Dr. No* to a high-

Bond battles an evil monkey

IQ monkey, perched on a henchman's shoulders. A later writer was so pessimistic about the final script (sans monkey) he insisted his name be removed. Half a dozen stars rejected the lead before the producers settled on a barely known 32-year-old named Sean Connery (prior movies: *Tarzan's Greatest Adventure, Darby O'Gill and the Little People*). Connery had driven milk trucks before acting. The distributor doubted they could sell the picture in major US cities because there was "a Limey truck driver playing the lead," so it opened the film at drive-in theaters in Oklahoma and Texas.

Developing the twenty-sixth Bond movie or the tenth statin is an entirely different experience. Actors compete for roles in Bond #26; studios line up for marketing rights; cash pours in. We now understand that a British spy wearing Brioni can sell tickets, just as we know that the next statin can lower cholesterol. There can be bumps in the road—Baycol (the sixth statin) was withdrawn due to unexpectedly high toxicities; Timothy Dalton (the fourth Bond) happened—but the directions are clear. Bond needs a bad guy, a fast car, a smooth drink, a double-crossing damsel, and a few double entendres. Follow-on drugs need to clear a known list of safety and efficacy hurdles. Franchise projects are easier to understand than loonshots, easier to quantify, and easier to sell up the chain of command in large companies. The challenge for these sequels and follow-ons is not in making it through the long dark tunnel of skepticism and uncertainty. Their challenge is in exceeding what came before.

Bond and the statins, of course, survived those challenges just fine. The Bond films became the most successful film franchise in history. The statins became the most successful drug franchise in history.

Inventors or creatives championing loonshots are often tempted to ridicule franchises—as Steve Jobs 1.0 did with the "bozos" developing Apple II follow-ons. But both sides need each other. Without the certainties of franchises, the high failure rates of loonshots would bankrupt companies and industries. Without fresh loonshots, franchise developers would shrivel and die. If we want more *Junos* and *Slumdog Millionaires*, we need the next *Avengers* and *Transformers.* If we want better drugs for cancer and Alzheimer's, we need the next statin.

Pixar, as Catmull and others have described, created an environment well known for what we would call phase separation and dynamic equilibrium, for nurturing loonshots while maintaining a balance between loonshots and franchises. But perhaps the most interesting lesson that was readily visible at Pixar, key to escaping the Moses Trap, was the difference between two ways of leading, which I'll call *system mindset* and *outcome mindset.*

For the clearest explanation of this difference, let's turn to a board game.

HOW TO WIN AT CHESS

Garry Kasparov reigned as world chess champion for fifteen years, the longest record in the history of the game. He ranks, on many lists, as the greatest chess player of all time. The difference between system and outcome mindset is a principle I adapted from his book *How Life Imitates Chess.* Kasparov describes this principle as key to his success.

We can think of analyzing *why* a move is bad—why pawn-takes-bishop, for example, lost the game—as level 1 strategy, or outcome mindset. After a bad move costs him a game, however, Kasparov analyzes not just why the move was bad, but how he should change the *decision process behind the move.* In other words, *how* he decided on that move, in that moment, in the context of that opponent, and what that means for how he should change his decision-making and game-preparation routine in the future.

Garry Kasparov

Analyzing the decision process behind a move I'll call level 2 strategy, or system mindset.

The principle applies broadly. You can analyze *why* an investment went south. The company's balance sheet was too weak, for example. That's outcome mindset. But you will gain much more from analyzing the *process* by which you arrived at the decision to invest. What's on your diligence list? How do you go through that list? Did something distract you or cause you to overlook or ignore that item on the list? What should you change about what's on your list or how you conduct your analyses or how you draw your conclusions—the *process behind the decision* to invest—to ensure that mistake won't happen again? That's system mindset.

You can analyze *why* you argued with your spouse. It was, let's say, your comment about your spouse's driving. But you may improve marital relations even more if you understand the *process* by which you decided it was a good idea to offer that comment. What state were you in and what were you thinking before you said it? Are there some different things you might do when you are in that state and think those thoughts? How good would it feel to sleep in your own bed?

Let's apply the same principle to organizations. The weakest teams don't analyze failures at all. They just keep going. That's zero strategy.

Teams with an outcome mindset, level 1, analyze why a project or strategy failed. The storyline was too predictable. The product did not stand out enough from competitors' products. The drug candidate's data package was too weak. Those teams commit to working harder on storyline or unique product features or a better data package in the future.

Teams with a system mindset, level 2, probe the decision-making process behind a failure. How did we arrive at that decision? Should a different mix of people be involved, or involved in a different way? Should we change how we analyze opportunities before making similar decisions in the future? How do the incentives we have in place affect our decision-making? Should those be changed?

System mindset means carefully examining the quality of decisions, not just the quality of outcomes. A failed outcome, for example, does not *necessarily* mean the decision or decision process behind it was bad. There are good decisions with bad outcomes. Those are intelligent risks, well taken, that didn't play out. For example, if a lottery is paying out at 100 to 1, but only three tickets are sold, one of which will win, then yes, purchasing one of those three tickets is a good decision. Even if you end up holding one of the two that did not win. Under those same conditions, you should always make that same decision.

Evaluating decisions and outcomes separately is equally important in the opposite case: bad decisions may occasionally result in good outcomes. You may have a flawed strategy, but your opponent made an unforced error, so you won anyway. You kicked the ball weakly toward the goalkeeper, but he slipped on some mud, and you scored. Which is why probing wins, critically, is as important, if not more so, as probing losses. Failing to analyze wins can reinforce a bad process or strategy. You may not be lucky next time. You don't want to be the person who makes a poor investment, gets lucky because of a bubble, concludes he is an investment genius, bets his fortune, and then loses it all next time around.

At Pixar, Catmull probed both systems and processes, after both wins and stumbles. How should the feedback process, for example, be adjusted so a director is given the most valuable possible input, in a form most likely to be well received? Artists tend to hate feedback from suits or marketers or anyone outside their species, but they welcome it from thoughtful peers.

So at Pixar, every director receives private feedback on their project from an advisory group of other directors—and, in turn, serves on similar groups for other directors. And more: How might a director's incentives distort their decision-making process on budgets, timelines, and quality? How should those distortions be countered? What filmmaking habits are in place, for outdated reasons, that might be unnecessary or counterproductive today?

Like Vannevar Bush, who insisted, as described in chapter 1, that he "made no technical contribution whatever to the war effort," Catmull saw his job as minding the *system* rather than managing the *projects*.

That message got through to Jobs. Jobs had a role in the system—he was a brilliant deal-maker and financier. It was Jobs, for example, who insisted on timing the Pixar IPO with the *Toy Story* release, and Jobs who negotiated the Pixar deals with Disney. But he was asked to stay out of the early feedback loop on films. The gravity of his presence could crush the delicate candor needed to nurture early-stage, fragile projects. On those occasions he was invited to help near-finished films, Jobs would preface his remarks: "I'm not a filmmaker. You can ignore everything I say." Jobs had learned to mind the system, not manage the project.

Relinquishing control of a creative project and trusting in the inventor or artist or any other loonshot champion is not the same as relinquishing attention to detail. The chief executive at Genentech for fourteen years, Art Levinson, was famous—and feared—for his insistence on scientific precision. A few years ago, at the largest annual biotech meeting, Levinson strode onto the stage to give his keynote presentation, pointed at the conference organizer's logo behind him—a giant image of a DNA helix—and said, "This is a left-handed helix. It doesn't exist in nature." (DNA molecules are right-handed helices.) The crowd roared. We are a major industry, he explained, we should get our DNA right. He sent a message that inspired every scientist in the room. Science matters. Precision matters.

I often heard stories from scientist and manager friends at Genentech about Levinson. How he would call a junior technician in the lab, for example, and grill him on his data. Levinson and the early founders of Genentech understood, like Bush and Vail, and Catmull decades later, the need to tailor the tools to the phase. Ferocious attention to scientific detail—or artistic vision or engineering design—is one tool, tailored to

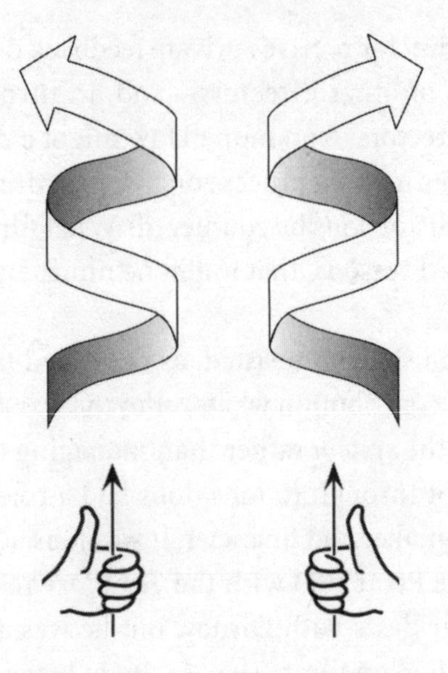

Left-handed vs. right-handed DNA

the phase, that motivates excellence among scientists, artists, or any type of creative.

Genentech achieved the highest levels of respect from the scientific community. It ranked behind only MIT in the number of citations per paper. It did so *without* sacrificing excellence in franchise. It not only developed four of the most important cancer drugs of the past twenty years but it also overcame the nearly impossible manufacturing challenges of growing them from live organisms, in a lab, to deliver to millions of patients around the world. The company translated that scientific and manufacturing expertise into products generating over $10 billion in annual sales. It did so, in large part, by balancing loonshots and franchises extraordinarily well.

In April 2000, three years after Steve Jobs returned to Apple, he invited Art Levinson to join his new board of directors. After Jobs passed away in 2011, Levinson replaced him as chairman of Apple.

RESCUE OPERATIONS

The well-told story of Jobs's return to Apple and its subsequent rise to the most valuable company in the world is a remarkable example of nurtur-

ing loonshots, in a race against time, to rescue a franchise in crisis. But it should be, by now, a familiar example.

Vannevar Bush arrived in Washington to rescue a franchise lagging far behind in a technology race, just months before the start of a world war. Bush's system helped create not only the dominant military force in the world (as we will see in chapter 8) but also the dominant national economy. Theodore Vail returned to AT&T to rescue a franchise in crisis after its telephone patent had expired and competitors were clawing at its lead. Vail's system not only transformed AT&T into the country's most successful business but also produced Nobel Prize–winning discoveries that launched the electronics age.

In Apple's case, the rescue operation began in December 1996, when Apple announced the acquisition of NeXT and Jobs's appointment as an advisor. It was the company's last-gasp attempt to save itself. Apple's operating system and machines were outdated. Three prior operating system overhauls, intended to compete with Microsoft Windows, had failed. Market share had plunged below 4 percent. Massive financial losses and heavy debts had pushed Apple to the edge of bankruptcy. The board tried, and failed, half a dozen times to find a buyer for the company. Elevating Jobs first to interim CEO in mid-1997 and then to full-time CEO in early 1998 was viewed as a Hail Mary play, and one with a particularly small chance of saving the company. The many failed promises of NeXT had reduced Jobs's credibility as a technology leader in the eyes of industry analysts and observers.

When Jobs finally took over, gone was the dismissive attitude toward soldiers. In March 1998, he hired Tim Cook, known as the "Attila the Hun of inventory," from Compaq to run operations.

Also gone were the blinders to S-type loonshots. For example, by 2001 music piracy on the internet was rampant. The idea of an online store selling what could easily be downloaded for free seemed absurd. And no one sold music online that customers could keep on their own computers (online music, at the time, was available only through subscription: monthly fees for streaming songs). Plus one more nutty thing: no one sold individual songs, at 99 cents each, rather than whole albums. "You're crazy," anyone could have told Jobs. "There's no way that could make any money."

The idea didn't seem so crazy after one million songs were downloaded

from the iTunes store in the first six days. There were no new technologies. Just a change in strategy that no one thought could work.

Apple's P-type loonshots, of course, transformed their industries: the iPod, the iPhone, and the iPad. But what ultimately made them so successful, aside from excellence in design and marketing (most, although not all, of the technologies inside had been invented by others), was an underlying S-type loonshot. It was a strategy that had been rejected by nearly all others in the industry: a closed ecosystem.

Many companies had tried, and failed, to impose a closed ecosystem on customers. IBM built a personal computer with a proprietary operating system called OS/2. Both the computer and the operating system disappeared. Analysts, observers, and industry experts concluded that a closed ecosystem could never work: customers wanted choice. Apple, while Jobs was exiled to NeXT, followed the advice of the analysts and experts. It opened its system, licensing out Macintosh software and architecture. Clones proliferated, just like Windows-based PCs.

When Jobs returned to Apple, he insisted that the board agree to shut down the clones. It cost Apple over $100 million to cancel existing contracts at a time when it was desperately fighting bankruptcy. But that S-type loonshot, closing the ecosystem, drove the phenomenal rise of Apple's products. The sex appeal of the new products lured customers in; the fence made it difficult to leave. Just like the failure of Friendster prior to Facebook, or the failure of cholesterol-lowering drugs and diets prior to Endo's statins, or the failure of the Comets before the Boeing 707, IBM's failure with OS/2 had been a False Fail.

In rescuing Apple, Jobs demonstrated how to escape the Moses Trap. He had learned to nurture both types of loonshots: P-type and S-type. He had separated his phases: the studio of Jony Ive, Apple's chief product designer, who reported only to Jobs, became "as off-limits as Los Alamos during the Manhattan Project." He had learned to love both artists and soldiers: it was Tim Cook who was groomed to succeed him as CEO. Jobs tailored the tools to the phase and balanced the tensions between new products and existing franchises in ways that have been described in many books and articles written about Apple. He had learned to be a gardener nurturing loonshots, rather than a Moses commanding them.

"The whole notion of how you build a company is fascinating," Jobs

told his biographer, Walter Isaacson. "I discovered that the best innovation is sometimes the company, the way you organize."

Jobs arrived at the same conclusion that the military historian James Phinney Baxter did half a century earlier, reflecting on the success of Bush's system in turning the course of World War II: "If a miracle had been accomplished anywhere along the line," Baxter wrote, "it was in the field of organization, where conditions had been created under which success was more likely to be achieved in time."

§

THE FIRST THREE RULES

The example of Xerox PARC has appeared several times above. Before pulling together what we've learned from the five stories in the five chapters of part one, it's worth briefly touching on what happened at PARC. It highlights another side—the inverse—of the Moses Trap.

In 1970, Xerox was a shining symbol of innovation. It was the first company, before Apple, to reach a billion dollars in sales in less than ten years from introducing a single technology—the photocopier. But by 1970, the photocopier franchise had matured, so the leaders of Xerox decided to create a separate unit, in Palo Alto, California, far away from its headquarters in New York and its manufacturing division in Texas, to explore new technologies. They called it the Palo Alto Research Center. PARC attracted the best and the brightest. Engineers at PARC would end up winning many of the most prestigious awards in computer science and founding, or joining and transforming, many of the earliest pioneering computer companies (including Apple).

During the 1970s, engineers at PARC invented the first graphics-enabled personal computer (the Alto), the first visual-based word processor, the first laser printer, the first local networking system (ethernet), the first object-oriented programming language, and a half dozen other firsts. It was an incredible run. But none of those breakthroughs was commercialized by Xerox.

"Some companies are the equivalent of an innovation landfill," wrote one senior Apple executive, who helped lure some of PARC's best engineers to Apple. "They are garbage dumps where great ideas go to die. At

PARC, the key development people kept leaving because they never saw their products get into the market."

One of the Alto project leaders at PARC gradually realized it was the company's "structure, not cost estimates or technological visions," that was driving apart the loonshot group at PARC and the franchise group in Texas, the one that made typewriters and other office machines. The Texas group "had to sandbag the Alto III, because with it they wouldn't make their numbers and therefore wouldn't get their bonuses. It would have been an absolutely impossible burden on them to be successful in making typewriters and also introduce the world's first personal computer. And they should never have been asked to do it that way. So it was shot down."

In other words, as mentioned earlier, the weak link is not the supply of ideas. It is the transfer to the field. And underlying that weak link is structure—the design of the system—rather than the people or the culture.

* * *

PARC was an example of the *opposite* of the Moses Trap. Phase separation succeeds brilliantly. But rather than loonshots commanded out of the group by a Moses, loonshots are ignored or actively quashed ("Color is not part of the office of the future") and never emerge.

PARC was far from the only example. The PARC Trap—loonshots stay parked, and never leave—is common. In 1975, for example, Steve Sasson at Kodak's research lab developed one of the earliest digital cameras. Kodak buried it for a decade.

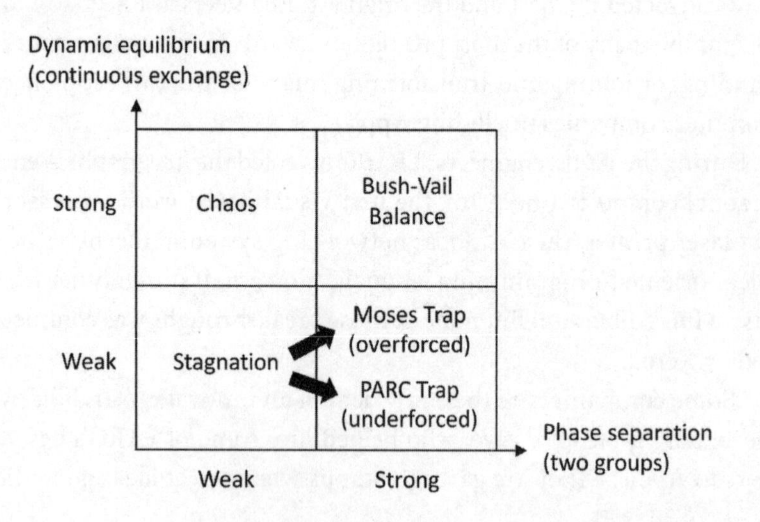

Well-intentioned leaders can create a high-performing, quarantined research group, as Xerox's leaders did, with an environment well suited to creativity and invention. The organization moves out of the bottom-left "stagnation" quadrant below, to the right. But there will always be resistance, in the franchise groups, to new ideas, just as the US military resisted, at first, many of the technologies emerging from Vannevar Bush's group.

Getting the touch and balance right requires a gentle helping hand to overcome internal barriers—the hand of a gardener rather than the staff of a Moses. If the transfer is either overforced (a thunderous commandment) or underforced (no helping hand), promising ideas and technologies will languish in the labs. The organization will lose the technologies, it will lose the race against time, and it will lose the loyalty of its inventors, who won't stay around for long.

The stories in part one illustrate the first three Bush-Vail rules:

1. **Separate the phases**
   - Separate your artists and soldiers
   - Tailor the tools to the phase
   - Watch your blind side: nurture both types of loonshots (product and strategy)

2. **Create dynamic equilibrium**
   - Love your artists and soldiers equally
   - Manage the transfer, not the technology: be a gardener, not a Moses
   - Appoint, and train, project champions to bridge the divide

3. **Spread a system mindset**
   - Keep asking *why* the organization made the choices that it did
   - Keep asking *how* the decision-making process can be improved
   - Identify teams with outcome mindsets, and help them adopt system mindsets

This book began with examples of wildly innovative groups that suddenly stopped innovating. Those examples included Catmull's description of a decline at Disney: "The drought that was beginning then [following *The Lion King*] would last for the next sixteen years. . . . I felt an urgency to understand the hidden factors that were behind it."

The Bush-Vail rules above describe *how* you can prevent the decline and stagnation that follows a phase transition, when good teams begin killing great ideas. But we have not yet gotten to the *what* and the *why*: what are those hidden forces and why do they appear? In other words, what causes that transition?

So now let's turn to the what and the why. Understanding those forces will reveal a fourth category of Bush-Vail rules.

We'll begin with a legendary detective and an equally famous political philosopher.

Both specialized in hidden forces.

PART TWO

THE SCIENCE OF SUDDEN CHANGE

While the individual man is an insoluble puzzle, in the aggregate he becomes a mathematical certainty.

—Sherlock Holmes, in *The Sign of the Four*

Interlude

The Importance of Being Emergent

Why should you believe any rule or generalization about teams or companies or any group of people? All people are different. All teams are different.

Yet some rules that describe what happens when many people come together to accomplish tasks seem to work pretty well. The rules of efficient markets, invisible hands, and so on. Those have been established and tested beyond any doubt, right?

Well, sort of. This is Alan Greenspan, the economist who chaired the US Federal Reserve for nineteen years, writing in the *Financial Times* in 2011:

> Today's competitive markets, whether we seek to recognise it or not, are driven by an international version of Adam Smith's "invisible hand" that is unredeemably opaque. With notably rare exceptions (2008, for example), the global "invisible hand" has created relatively stable exchange rates, interest rates, prices, and wage rates.

Here's the problem: analyzing markets except for the "notably rare exceptions" of bubbles and crashes is like analyzing the weather except for storms and droughts. We really *do* want to understand storms and droughts. We'd like to know if we will need an umbrella.

Not all economists, to be fair, agreed with Greenspan. One extended his logic to the analysis of diplomacy: "With notably rare exceptions, Germany remained largely at peace with its neighbors during the 20th century."

Greenspan's view, however, that efficient markets and invisible hands are fundamental laws that are rarely, if ever, violated is widespread. But it's a fallacy. That fallacy is a common cause of policy disasters (or investment opportunities, if you are a trader).

Neither efficient markets nor invisible hands are fundamental laws. They are both *emergent properties*.

Emergent properties are collective behaviors: dynamics of the whole that don't depend on the details of the parts, the macro that rises above the micro. Molecules will flow at high temperatures and freeze at low temperatures regardless of the differences in their details. A water molecule has three atoms and is shaped like a triangle. Ammonia has four atoms and is shaped like a pyramid. Molecules of buckminsterfullerene have sixty atoms and are shaped like soccer balls (they're called buckyballs for short). Yet they *all* exhibit the *same* fluid dynamics at high temperatures and solid dynamics at low temperatures.

One of the things that distinguishes an emergent property like the flow of liquids from a fundamental law—like quantum mechanics or gravity, for example—is that an emergent property can suddenly change. With a small shift in temperature, liquids suddenly change into solids. That sudden shift from one emergent behavior to another is *exactly* what we mean by a phase transition.

Although all people are different, and all teams are different, what makes emergent properties and the phase transitions between them so interesting is that they are so predictable. We will see why organizations will *always* transform above a certain size, just like water will *always* freeze below a certain temperature, traffic will *always* jam above a critical density of cars, and one burning tree in a forest will *always* explode into a wildfire in high winds. These are all examples of phase transitions.

Each person and team may be a puzzle. But in the aggregate, as Sherlock Holmes might say, the likelihood that any group will experience a phase transition becomes a mathematical certainty.

The terrific thing about the science of emergence is that once we understand a phase transition, we can begin to *manage* it. We can design stronger materials, build better highways, create safer forests—and engineer more innovative teams and companies.

***

So what does all of this tell us about Mr. Greenspan and the widespread belief in the almighty invisible hand? The confidence in the infallibility of the invisible hand is a consequence, to come back to our Newton-Jobs theme from the prior chapter, of false idolatry. For two hundred years we have been bowing down to the *wrong* seventeenth-century physicist.

To see what I mean, let's travel back to a summer day in Britain two centuries ago.

On Sunday, July 11, 1790, as he lay dying in his home in Edinburgh, a revered Scottish philosopher, who would become famous for ideas he did not believe and a phrase he did not invent, sent for two friends. He begged them to burn his unpublished notes and manuscripts, except for one. The two had been resisting similar requests for months, hoping he would change his mind. That Sunday they yielded. They burned, in total, sixteen volumes. The scholar, relieved, joined his friends for supper. At half past nine, he rose to return to bed, announcing, "I love your company, gentlemen, but I believe I must leave you to go to another world." Six days later he died.

Adam Smith, who knew how to make an exit, has grown into a misty icon, a hero to libertarians and free-marketeers who like their economics neat, hold the morals. (The real Adam Smith argued for restraints on markets and prized his work on ethics more than his work on economics.) The manuscript Smith asked his friends to spare had nothing to do with either ethics or economics. It was his *History of Astronomy*, written shortly after he finished graduate school.

In the *History*, Smith states that the task of the philosopher is to explain "the connecting principles of nature . . . the invisible chains which bind together" disjointed observations. Smith analyzes competing theories of planetary motion and ends with a deep bow to Newton, whose theory of gravity he describes as "the greatest discovery that ever was made by man." (Newton worship was all the rage at the time. For a taste, see the wonderfully titled *Sir Isaac Newton's Philosophy Explain'd for the Use of the Ladies.*)

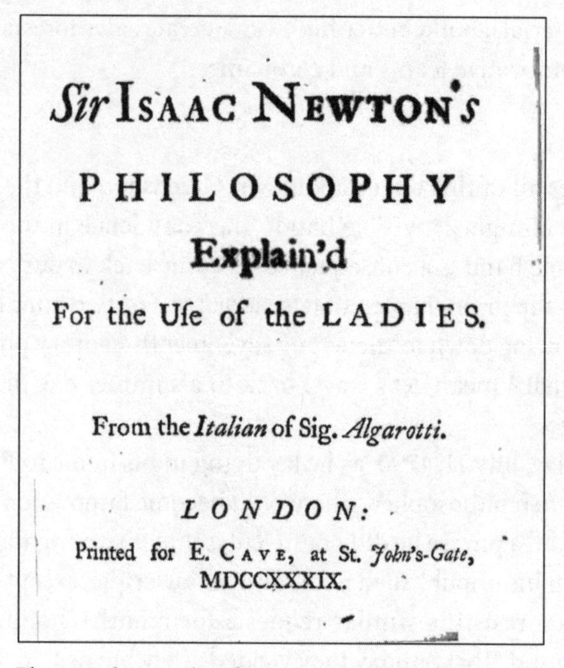
Sir ISAAC NEWTON's

PHILOSOPHY

Explain'd

For the Uſe of the LADIES.

From the *Italian* of Sig. *Algarotti.*

LONDON:

Printed for E. CAVE, at St. *John's-Gate,*

MDCCXXXIX.

The complete works of Isaac Newton (abridged), 1739

The idea of an underlying force that can explain complex behaviors, as gravity explained the motion of planets and tides, fascinated Smith. His *Theory of Moral Sentiments* (1759) proposes an underlying force that explains how humans behave. His *Wealth of Nations* (1776) proposes an underlying force that explains how markets behave.

Smith didn't intend to call that underlying force in markets an *invisible hand.* He used the phrase only three times, across all his writings, ambiguously and inconsistently. (The first time, he used it as a sarcastic put-down of superstitious beliefs, a "mildly ironic joke.") The hand metaphor had been used by many writers, and it was ignored, in the context of financial markets, for 170 years after Smith died, until an economics textbook in the 1950s revived the phrase, imbued it with its current meaning, and attributed that meaning retroactively to Smith.

Whatever its origin, the *current* meaning has become widely accepted as fact: individuals acting purely out of self-interest can create complex market behaviors. Prices will adjust to demand, resources will be allocated efficiently, and so on. Shopkeepers sell, people shop, and these collective behaviors just . . . emerge. The same behaviors appear whether butchers sell chicken or beef, whether bakers sell cupcakes or bread. They are dynamics of the whole that don't depend on details of the parts.

That should sound familiar. Liquids will flow the same way whether they are made of water or ammonia. The collective behavior of markets is an emergent property like the flow of liquids, *not* a fundamental law like gravity. For two centuries economists have aspired to Newton-style fundamental laws (the "Gravity Model" of international trade, a "Quantum Theory Model of Economics," "Conservation Laws" of economics—all by Nobel laureates). These economists have been inspired by Newton, who grew into the high priest of one branch of physics; call it Physics Catholicism. That branch of physics preaches a dogmatic belief in fundamental laws and the glamorous search to discover them.

Adam Smith's work, however, was much closer to the field's quieter, and less well-known, Protestant offshoot: the study of emergent phenomena. The high priest of that offshoot was Newton's widely admired contemporary, Robert Boyle.

The battles between the descendants of these two branches continue to this day. One side believes the highest priority should be to search for fundamental laws and writes lines like: "We are living through a landmark period in human history in which the search for the ultimate laws of the universe will finally draw to a close."

The other side believes that there may be no such fundamental laws. The laws of nature may be like an infinite skyscraper, with different and fascinating rules at each level, rules that are gradually revealed as you descend the stairway down to smaller and smaller distance scales. The current high priest of this branch writes lines like: "The existence of these [emergent] principles is so obvious that it is a cliché not discussed in polite company," and of their deniers, "The safety that comes from acknowledging only the facts one likes is fundamentally incompatible with science. Sooner or later it must be swept away by the forces of history." (The author of those latter two quotes, Bob Laughlin, is a Nobel laureate and disciple of Phil Anderson, mentioned earlier. He required all his students, of which I was one, to read Anderson's "More Is Different" essay.)

It may seem like the distinction between dueling branches of physics should be of little interest to a nonspecialist, just like the distinction between dueling branches of religion may be of little interest to an atheist. But the distinction can matter, a lot. Perfectly efficient markets—a

Loonshot Widely dismissed or ridiculed idea		Franchise
1922	A 12-year-old patient with diabetes is treated with ground-up pancreas extract	Insulin
1935	An 80-pound payload is accelerated to 500 miles per hour through rocket propulsion	Long-range ballistic missiles
1961	A 32-year-old former milk-truck driver plays a metrosexual British spy who saves the world	James Bond
1976	A script titled *The Adventures of Luke Starkiller* is green-lit	*Star Wars*

Phase transition Sudden transformation in system behavior, as one or more control parameters cross a critical threshold	
Water	From liquid to solid, as temperature decreases
Cars on highways	From smooth flow to jammed flow, as car density increases
Fires in forests	From contained to uncontrolled, as wind speed increases
Individuals in companies	From a focus on loonshots to a focus on careers, as size of company increases

Newton-style, fundamental belief—don't have bubbles and crashes. Boyle-style emergent markets, on the other hand, with certain reasonable assumptions, almost always do.

Which brings us to the importance of being emergent—or, at least, understanding emergence. It can help us capture the benefits of diversity while reducing the risk of collective disasters. We want to benefit from the wisdom of crowds while reducing the risk of market crashes. We want to benefit from a plurality of beliefs while reducing the risk of religious wars.

Over the coming chapters, we will apply Boyle-style science to help us understand the collective behaviors of individuals in companies, much as Smith applied Boyle-style science (not Newton-style) to help us understand the collective behaviors of individuals in markets.

Understanding these behaviors will help us learn what we really want to know: how to capture the benefits that large groups bring to big goals—winning wars, curing diseases, transforming industries—while reducing the risks that those groups will crush valuable and fragile loonshots.

To see how this works, let's start with a drive along the highway.

6

Phase Transitions, I: Marriage, Forest Fires, and Terrorists

When gradual shifts cause sudden transformations

You're driving home from work, on the highway, you're anxious, maybe speeding a bit, but the traffic is flowing well. Suddenly, the highway turns into a parking lot of stopped cars. There's no visible cause. There are no on-ramps or accidents in sight. You set aside thoughts of a cold dinner and angry spouse and wonder: where did this traffic jam come from?

Answer: You have just experienced a phase transition—a sudden change between two emergent behaviors. Those two behaviors are *smooth flow* and *jammed flow.*

Here's how to think about it: Imagine the highway is nearly empty. The driver of the car ahead of you, hundreds of yards away, briefly taps then releases his brakes—maybe he's seen a squirrel. You see the red brake lights briefly flash, but since the car is so far away, there is no need to slow down.

On a crowded highway, the same driver is only a few car lengths ahead. As soon as he touches his brakes, you slam on yours. The brake lights in front of you may have only flashed for two seconds. But once the driver ahead of you releases his brakes, and you release yours, it takes you longer than two seconds to accelerate back to cruising speed. It may take you four seconds. The delay grows for the driver behind you. It may take him eight seconds to get back to the original speed. For the driver behind him, 16 seconds. The one small tap grows exponentially, until it becomes a traffic jam.

In the early 1990s, a pair of physicists showed that below a critical density of cars on the highway, traffic flow is stable. Small disruptions—drivers tapping their brakes when squirrels run by—have no effect. Traffic engineers call that a *smooth flow* state. But above that threshold, traffic flow suddenly becomes unstable. Small disruptions grow exponentially. That's a *jammed flow* state. The sudden change between smooth and jammed flow is a phase transition.

As rush hour nears, the density of cars is right on the verge of that critical threshold. A few extra cars on some stretch of the highway—a pileup, for example, behind a slow-moving truck—will push traffic flow over the edge.

The stalls that mysteriously appear with no apparent cause are called *phantom jams*. They have been confirmed not only by careful observation on highways but by experiment. In 2013, a group of researchers in Japan tracked cars circling inside the Nagoya Dome, an indoor baseball field. They found, as predicted, that when the density of cars exceeded a critical threshold, spontaneous jams suddenly appeared.

Over the past two decades, traffic flow researchers have introduced many variations on the basic model introduced in the 1990s: more vs. less aggressive drivers, faster vs. slower reaction times, a mix of big cars (trucks) and small cars, and so on. In all cases, they find the same phase

Testing the traffic flow phase transition at the Nagoya Dome in Japan

transition. When the density exceeds a critical threshold, the system will flip from the smooth-flow to the jammed-flow state.

Phase transitions are everywhere.

* * *

To understand what phase transitions tell us about nurturing loonshots more effectively, we need to know just two things about them:

1. At the heart of every phase transition is a tug-of-war between two competing forces.
2. Phase transitions are triggered when small shifts in system properties—for example, density or temperature—cause the balance between those two forces to change.

That's it.

To illustrate these two ideas, we'll set aside for a moment traffic flow, which can be complicated, and start with something much simpler: marriage.

JANE AUSTEN, PHYSICIST

> It is a truth universally acknowledged that a single man in possession of a good fortune must be in want of a wife.
>
> —Jane Austen, *Pride and Prejudice*

Miss Austen suggests that two competing forces tug at single men. Those of modest fortunes, in their younger and more aggressive years, may travel widely in the pursuit of fame, wealth, and glory. Let's call that force "entropy."

Those of greater fortune, in their older and gentler years, want to settle down with a partner. They seek family, stability, and cable TV. Let's call that force "binding energy."

The physicist Richard Feynman once said, "Learn by trying to understand simple things in terms of other ideas—always honestly and directly." His disciple Lenny Susskind, my former graduate advisor, took that advice seriously. Lenny once explained to me a complex idea in topology, the study of surfaces, by saying, "Imagine an elephant, then take its trunk and shove it up its a—. That's your surface."

In that vivid spirit, imagine the bottom half of a very large egg carton. To be specific, imagine a square carton, 20 by 20 egg wells, so there are 400 wells in total. Let's seal our carton inside a glass protective cover, so we can peer inside and inspect the wells. Rather than imagine eggs, since we will be doing a lot of jiggling, and that could get messy, let's visualize Jane Austen's men as small marbles resting in those egg wells. Those marbles have settled down. They are happily married, raising kids.

Now imagine gently shaking the egg carton back and forth. The marbles rock within their small egg wells. But they stay put. Now gradually increase the vigor of your shaking. The marbles reach higher and higher up the sides of their wells. But still they stay put. Finally, when the vigor of your shaking crosses a certain critical threshold, so that a marble reaches the top of its well—all hell breaks loose. Marbles leave their wells and end up in their neighbor's well; they quickly leave that one and go on to the next; they bounce into and off of other marbles; they travel everywhere, all over the place. Rather than rest quietly in an ordered pattern, the marbles randomly ricochet around the carton, creating a scattered, disordered sea of marbles.

Welcome to a Manhattan singles bar.

In physics language, we triggered a marble-solid to marble-liquid phase transition.

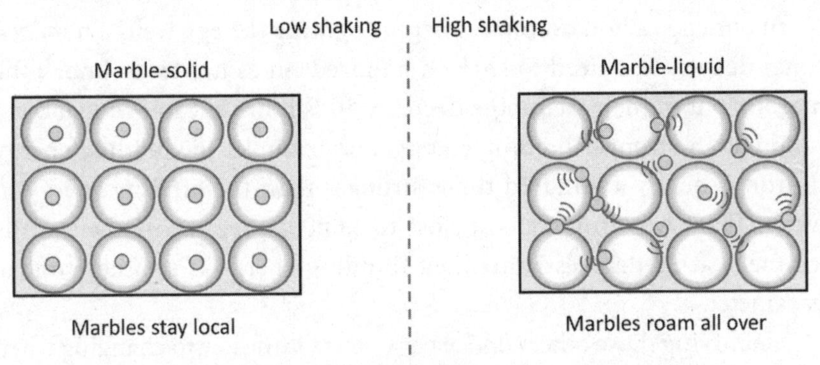

The marble-solid to marble-liquid phase transition: when shaking energy crosses a threshold, marbles suddenly break free

The system property that we gradually change to trigger a phase transition is called a *control parameter.* In the traffic flow example, the density of cars on the highway is the control parameter. In this marble-solid to marble-liquid transition, the vigor of our shaking is the control parameter. Shaking vigor can be measured on a scale. We can call that scale "temperature." The hotter the temperature, the more the entropy term dominates (the urge to roam all over). The colder the temperature, the more the binding energy dominates (the attraction to the bottom of the egg well). When the temperature crosses a threshold—a breakeven point between entropy and binding energy—the system suddenly changes behavior. That's a phase transition.

In real solids, the binding energy arises from the forces between molecules rather than a fixed landscape of small wells. But otherwise the model gets it right. This microscopic tug-of-war between entropy and binding energy is behind every liquid-to-solid phase transition.

* * *

In the next chapter, I will show you that team size plays the same role in organizations that temperature does for liquids and solids. As team size crosses a "magic number," the balance of incentives shifts from encouraging a focus on loonshots to a focus on careers.

The magic number is not universal, however. Teams transform at different sizes, just like solids melt at different temperatures. The reason is the key idea behind our fourth rule. It's why we can *change* the magic number. Systems have more than one control parameter.

In our egg carton example, imagine making the egg wells a hundred times deeper. You need to shake a hundred times harder to knock the marbles out of their wells. The deeper well is how we can think about a solid with a stronger binding energy. For example, the binding energy in iron is nearly a hundred times stronger than the binding energy in water. That's why iron melts at close to 2,800 degrees Fahrenheit, while ice melts at 32 degrees Fahrenheit. Binding energy is another control parameter.

Identifying those other control parameters is the key to changing when systems will snap: when solids will melt, when traffic will jam, or when teams will begin rejecting loonshots.

PHASE DIAGRAMS

Let's come back to our traffic flow example, and to a useful technique that scientists use to help think about these questions.

The two competing forces for drivers on a highway are speed and safety. A driver accelerates to reach cruising speed, but he brakes to avoid hitting the bumper in front of him. The spacing between cars—average car density—is one control parameter, as we saw above. But it's not the only one. Your decision whether to slam on your brakes when you see brake lights flash on the car ahead depends not only on the distance to that car but on how fast you're going. At 30 miles per hour, your stopping distance is roughly six car lengths. At 80 miles per hour, it's closer to 30 car lengths. In deciding whether to brake, your brain intuitively estimates your stopping distance and compares it with the distance to the bumper in front of you. *Both* the average speed of cars and the average density of cars contribute to triggering the transition.

A *phase diagram* captures these two control parameters in one graphic. In the diagram below, the average distance between cars is measured on the vertical axis, and the average car speed is measured on the horizontal axis. At low speeds or when there are few cars on the highway, above and to the left of the dashed line (#1 in the diagram below), traffic flows smoothly.

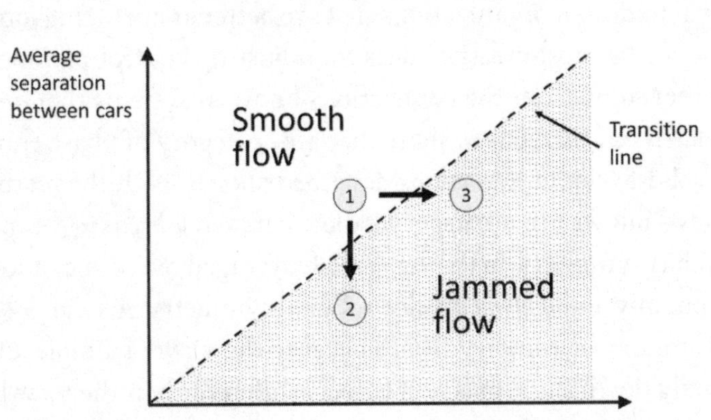

When the average separation between cars falls below the transition line (1 → 2) or the average speed rises above it (1 → 3), smooth flow suddenly jams

When the traffic flow crosses the transition line at either higher car densities (#2) or faster car speeds (#3), small disruptions grow exponentially into a jam. The dashed transition line slopes up and to the right because braking distances get longer as cars travel faster, meaning a larger average separation between cars is required to avoid a jam.

Traffic engineers use these ideas to design better highways. *Reducing speed limits* in heavy traffic may seem counterintuitive, but it reduces the likelihood that a small disruption will cause a jam (it can shift the flow from point #3 in the diagram above to #1). Some highways use *ramp metering*: when the density and speed start to approach the phase-transition dashed line in the figure above, on-ramp traffic signals can temporarily reduce the flow of new cars onto the highway. That backs the highway flow away from the dashed line. A policy of banning trucks from passing other trucks (called a *truck-overtaking ban*) reduces the pileups behind trucks. Those pileups temporarily increase the density of cars and can push smooth traffic flow across the dashed line and into a jam. Studies on German autobahns have shown that truck-overtaking bans work. They improve the flow of passenger vehicles, although they slightly decrease the flow of trucks.

* * *

The science of phase transitions, as we can see from the traffic flow example, has expanded far beyond an academic curiosity. Identifying the control parameters of a transition helps us *manage* that transition. Which is exactly what we will do with teams and companies: identify what we can adjust to design organizations that are better at nurturing loonshots.

Some of the most creative ideas for adjusting control parameters, as we will see, come from the connections between systems that *appear* to be unrelated but turn out to share the same category of phase transition.

The solid-to-liquid transitions described above—both the marbles and real solids—fall within a category called *symmetry-breaking* transitions. A liquid has symmetry in the sense that, averaged over time, it looks the same from any angle. That's called rotation symmetry. A solid does not: it "breaks" rotation symmetry. That's because the view of a molecule looking directly down the x-axis will be very different than the view looking five or ten degrees off that axis. Over a dozen Nobel Prizes have been awarded for discoveries that were ultimately explained by this same principle of a symmetry-breaking transition.

The sudden change in traffic flow falls within a second category of phase transition called a *dynamic instability.* A change in control parameters transforms one kind of motion (smoothly flowing cars) into a different kind of motion (jammed flow) by making the smooth flow very sensitive to small disruptions (driver tapping on his brakes). Fluids and gases also experience dynamic instabilities. They flow smoothly, but only at speeds below a critical threshold. Above that threshold, the flow suddenly becomes turbulent.

Imagine, for example, a boat moving slowly down a river. Water parts smoothly at the front of the boat. In the back, as water speeds up to fill in the space left behind, the flow forms a big, messy, turbulent wake. Or picture smoke rising from a cigarette in still air. In the picture below, the smoke from Bogart's cigarette breaks apart a few inches above the tip. Initially the smoke flows in a smooth column; as the smoke particles gather speed (the hot air coming off the tip of the cigarette accelerates upward), the column will suddenly break apart into a turbulent mess. Both the flow

Humphrey Bogart demonstrates the transition from smooth flow to turbulent flow

of water around a boat and cigarette smoke rising through the air are examples of the transition to turbulence. Because turbulence is closely connected to drag forces, understanding this kind of transition helps us design better ships, planes, and even golf balls. (Golf balls are dimpled because a little turbulence near a surface layer reduces drag—which is why, if you have a great swing, you can drive a modern dimpled golf ball over four hundred yards. Smooth golf balls would travel roughly half that distance.)

* * *

In 1957, a pair of British mathematicians identified a new category of phase transition. It has helped us understand the spread of forest fires, predict the formation of oil deposits, and, most recently, anticipate, and possibly prevent, terror attacks by analyzing the online behavior of would-be terrorists. Thanks to the magic of emergent behavior—of "more is different"—we now have a terror-hunting tool that can be used *without* violating online privacy.

It all began with a gas-mask puzzle.

FROM GAS MASKS TO FOREST FIRES

In 1954, a mathematician named John Hammersley presented an unusual paper at a meeting held at the offices of the Royal Statistical Society in London. He described new statistical techniques for evaluating the likelihood that certain patterns were due purely to chance.

Hammersley presented the example of Neolithic stone circles in western Scotland. The circles, built by Druids over three thousand years ago, measured from nine to one hundred feet in diameter. An engineer named Alexander Thom had studied the circles and claimed that each of them had been built to a multiple of a certain unit length. Professional archaeologists scoffed. One audience member described it as a controversy over whether we should think of Neolithic man as a savage or a colleague. But Hammersley's statistical methods supported Thom's claims. The Druids were more sophisticated than anyone had previously believed. They were indeed colleagues.

In the audience that day, Simon Broadbent, a 26-year-old engineer who published poetry on the side, was intrigued. Broadbent worked for the British Coal Utilisation Association analyzing coal production. He'd been asked to look into how to design better gas masks for coal miners. Gas masks use materials filled with pores small enough and sticky enough to trap dangerous particles as air passes through. The pores in those materials are of random sizes and randomly distributed. For a gas mask to work, those random pores must create at least one *connected* channel that allows air to flow all the way through the mask—from one side to the other, uninterrupted, so that a miner can breathe.

During the discussion session after the paper, Broadbent asked Hammersley if his techniques for analyzing randomness in data could predict which materials with random pores would contain at least one connected channel. In other words, if told the type of material, could Hammersley predict whether a coal miner wearing a mask made of that material would suffocate?

Hammersley soon realized that no one had ever posed, or at least answered, a statistical problem of that sort. The two began collaborating. Hammersley, 34, was a kind of odd-jobs statistician at Oxford (the field had not yet developed into a major discipline). His job was to tackle whatever problems the university administration or faculty suggested. One year he was asked to teach a course in the Department of Forestry on how researchers should collect and analyze data on tree growth. It wasn't long before Hammersley realized that Broadbent's question applied much more broadly than just to gas-mask design. It applied to forests as well.

Imagine a forest as a random distribution of trees. Now suppose a fire is started on one side of the forest. Assume fire can spread to a neighboring tree only if that tree is close enough for a spark to jump across. Will the fire spread from one edge of the forest to the other?

Broadbent and Hammersley discovered that the answer to both the gas-mask puzzle and the forest-fire puzzle was described by a phase transition. Below a threshold density of pores in a gas mask, no air could get through. Above that critical density, a channel would always appear connecting one side to the other. For forests, below a threshold density of trees, the fire would die out. Above that critical density, the fire would engulf the whole forest.

But tree density is not the only control parameter. Just as with cars on a highway, the forest-fire transition has more than one. Suppose wind is blowing strongly. Realistically, sparks could spread farther than just one tree. At high wind speeds, therefore, the contagion threshold should take

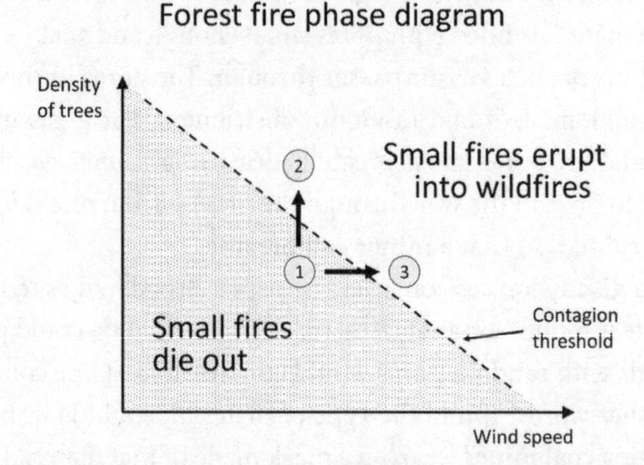

When the density of trees exceeds the contagion threshold (1 → 2) or wind speed crosses the same threshold (1 → 3), small fires will erupt into wildfires

place at a lower density of trees. In other words, the dashed transition line in the phase diagram above should slope downward to the right.

Air finding a channel through pores in a mask or fire finding a path through trees in a forest reminded Hammersley of water percolating through coffee grounds. If the grounds are packed too tightly, water may not find a path through. When they are loose enough: drip, drip. So Hammersley called his techniques and ideas "percolation theory."

Like symmetry-breaking, percolation theory turns out to connect a staggering range of seemingly unrelated systems.

When do rocks break? Rocks accumulate a random collection of stresses and fractures over time. When those small fractures coalesce into one large fracture that travels from one edge of a rock to the other, the rock breaks in two. That's the percolation threshold.

When should you drill for oil? Fissures deep in the ground form randomly, like pores in the gas mask. Below the percolation threshold for those fissures, your drill will likely hit a small, disconnected cluster of trapped oil. Bad investment. Above the percolation threshold, your drill is likely to pierce one giant, connected reservoir of oil. Good investment.

When will a small disease outbreak grow into an epidemic? Go back to the model of fire spreading from tree to tree. A high wind speed in the forest, blowing sparks quickly from tree to tree, is like a virus that is highly contagious. A high density of trees is like people living close together (in cities, for example). When the infectability and density cross a critical threshold, small outbreaks erupt into epidemics. When they fall below that threshold, small outbreaks die out quickly. That's the epidemic phase transition.

So how did real fire researchers react to these new mathematical models?

Not very well. It took a long time for firefighters to, um, warm up to statistical physicists and for the ideas to catch. Here's a story from a widely used fire-management textbook:

> The old-timer among firefighters [often] fails to realize how much can remain unknown to the man who has never had the opportunity to observe similar events personally.
>
> To demonstrate this lack of knowledge due to lack of experience, a middle-aged man tells a little story about himself of an event that happened more than 20 years ago. That day, the young fellow easily reached an observation point near the fire before an older ranger arrived. There before him was a rolling inferno of flames such as he had never before seen. Fascinated and frightened, he told himself that all the power of Man could never stop this fire.
>
> The old ranger wheezed up, rolled himself a cigarette and mumbled to himself, "The head will run into that old burn in half an hour and by sundown the wind will die and we'll cold trail her." Then he turned slowly to a messenger and said, "Joe, go phone headquarters and tell them the fire is under control."

These are not the kind of guys who will go gaga for differential equations.

HOW TO BE SIMPLE

In the 1990s, a handful of research groups finally succeeded in igniting interest in practical uses of percolation. For decades, forestry agencies had been using fire simulation models that captured the *micro*: the combustion properties of silvertop ash vs. ponderosa pine, the rate of fire spread as a function of slope steepness, and so on. Those models are helpful for

predicting hour-by-hour local behavior of a fire. Would it head left or right, speed up or slow down? But those models can't help with global patterns, the *macro*: the frequency, for example, of large fires.

To capture the interest of guys like the young man and old ranger in the story above, a research group composed of geologists, landscape ecologists, and physicists, found a middle ground between micro and macro. How they did that is key to what we will do in the next chapter with teams and companies.

Early forest-fire models didn't interest experienced firefighters because they were *too* macro, too simplistic. For example, they assumed that trees regrow everywhere in a forest at equal rates. They don't. Burned areas take decades to recover. The models also assumed that burning trees always ignite their neighbors. But in real forests, many things affect the spread of fires: air moisture, ground moisture, tree species, slope of the land. A fire will spread twice as fast, for example, on a 30 percent upward slope. Small burns nearly always spin out of control when humidity falls below 25 percent. But recording *all* those micro details, for every forest, would make predicting macro patterns impossible.

The researchers found a middle ground by creating a model that was *simple, but not simplistic.* Throw away too much detail, and you explain nothing. Retain all the detail—same thing. Do we need to know the difference between the combustion properties of silvertop ash and ponderosa pine to tease out general principles for designing safer forests? No. Will we need to sift through 137 case examples and dozens of theories to tease out general principles for designing more innovative teams and companies? No. We want a model that is *just simple enough* so that we can extract macro insights with confidence in their micro origins.

In other words, we want a model that describes the forest but is built from the trees.

To understand macro patterns of forest fires, it turns out, you need just two key parameters. I labeled the horizontal axis in the forest fire phase diagram a few pages before "wind speed." But a better term, which captures what really matters for the spread of fires, might be "virality." High wind speeds, dry ground, and low humidity increase virality: they make fires more likely to spread. Low wind speeds, moist ground, and high humidity decrease virality: they make fires less likely to spread.

In 1988, a fire in Yellowstone National Park burned 800,000 acres, 36 percent of the total park area—the largest fire in the park's history. Analyzing park policy is where percolation theory first showed what it can do. Until 1972, Yellowstone policy required rangers to put out every small fire immediately, whether it was caused by humans (a carelessly tossed cigarette) or by nature (a lightning strike). The frequency of small fires in a forest is sometimes called the *sparking rate*. The park managers' policy of reducing the sparking rate, although well intentioned, had allowed the forest to grow dense with old trees. They had inadvertently pushed the forest across the dashed line in the diagram above. Their policy had made contagion—a massive outbreak like the 1988 fire—inevitable.

Today most forestry services recognize the "Yellowstone effect" of artificially low sparking rates. They allow small- or medium-sized fires to burn under watch, called a controlled-burn policy. In some cases, if the forest is getting too close to the contagion threshold (the dashed line in the phase diagram), fire managers will initiate small burns, called prescribed burns, to back the forest away from the threshold.

The idea of a controlled burn seems sensible today, almost intuitive.

Percolation models helped spread that intuition by grounding the idea in science. But the most interesting success of those models—which led to a completely unexpected spin-off—came from comparing their predictions with historical records on the frequency of fires of different sizes.

The percolation models predict something you would *never* guess through intuition, or experience, or microsimulations with different tree types and vegetation. It is a unique prediction of the science of emergence and of phase transitions. According to these models, as a forest gets dangerously close to a phase transition, to erupting, the *frequency* of fires should take a specific form. The frequency should vary in inverse proportion to size: Twenty-acre fires should occur half as often as ten-acre fires. Forty-acre fires should occur one-quarter as often as ten-acre fires. Hundred-acre fires should occur one-tenth as often, and so on. That pattern, called a power law, is a surprising prediction—a mathematical clue that a forest is on the verge of erupting.

The pattern has been seen elsewhere. As we will discuss below, the power-law pattern is seen not only in forest-fire models, but in financial markets and terrorist attacks.

It would take another decade, however, for these three seemingly unrelated systems to come together. Outside of the forest-fire world, interest in Hammersley and Broadbent's percolation theory began to dwindle. Mathematicians explored variations on the puzzle: placing trees on the nodes of a square network (four neighbors per node), a hexagonal network (like the pattern on a soccer ball; three neighbors per node), cubic networks in 19 dimensions (38 neighbors), then trying to figure out at what density of trees a fire would erupt. After dozens of such variations had been analyzed and the big questions mostly answered, the theory gradually drifted into distinguished old age. It played quiet games of checkers with other elderly theories, rarely visited by the young.

The surprising rebirth of percolation theory began in January 1996. Four decades after Simon Broadbent asked John Hammersley an odd question about gas masks, a young Australian named Duncan Watts asked a math professor named Steven Strogatz an odd question about crickets.

SIX DEGREES OF KEVIN CRICKET

In the mid-nineties, Watts, a 24-year-old, six-foot-two-inch graduate of the Australian Defence Force Academy and part-time rock-climbing instructor, was a restless graduate student studying mathematics at Cornell University, growing bored of standard graduate school fare. He had been searching for a suitable thesis advisor when he came across Strogatz, 36, who had recently joined Cornell's applied math faculty. Strogatz specialized in quirky applications of advanced mathematical techniques (he once wrote a paper on the mathematics of *Romeo and Juliet*). At the time, Strogatz was working on understanding synchrony in nature: How do millions of heart cells beat in rhythm? How do thousands of fireflies flash at the same time? Watts was intrigued, and signed on as his student. After casting about for a problem to work on together, they settled on an insect puzzle: How do giant fields of crickets synchronize their chirping?

Watts and Strogatz began by collecting crickets and placing them in tiny individual soundproof boxes in a lab, each with built-in microphones and speakers. The idea was to play sounds of other crickets through the speakers. Adjusting who heard whom could test theories of synchronization.

How do crickets harmonize?

As Watts scrambled around the campus orchards collecting crickets, he wondered how connections formed between crickets in the wild, outside of his miniature cricket recording studio. Did crickets listen to their nearest neighbors? Did they listen to all neighbors closer than some distance? Was there a lead cricket conductor?

A Broadway play, *Six Degrees of Separation*, had recently popularized the idea that everyone was only a few friendships away from everyone else in society. Three college kids started a game called "Six Degrees of Kevin Bacon," ranking movie actors based on the same idea: one degree if you had been in a movie with Bacon, two degrees if you had been in a movie with someone who had been in a movie with Bacon, and so on. An astonishing 1.9 *million* actors are linked to Bacon by three degrees or less. What would "Six Degrees of Kevin Cricket" show?

As with Simon Broadbent's question about gas masks, Watts's question about crickets opened the door to a much bigger question. All sorts of networks had been explored for the percolation problem, as mentioned earlier. Square networks, hexagonal networks, networks in higher dimensions. But what about a social network? Where friends (crickets or humans or otherwise) could friend others, far removed?

The earlier percolation models made sense for studying the spread of fire, or an infectious disease, between objects that don't move, like trees in a forest. But crickets, quite famously, jump around. As do humans. You don't stay at home interacting only with neighbors living immediately to the left, right, front, and back of you. Over the course of a day, you might stop to chat with other parents as you drop off your kids at school. In the office, you might gossip about news or sports with colleagues at desks near yours or at the water cooler. At the grocery store, coming home from work, you might run into some friends and stop to catch up. And occasionally, during the day, or maybe a few times a week, you might reach out to connect with a friend across the country. That friend travels in a very different daily circle than yours.

The pattern of many connections within one tight community, punctuated by occasional ties to distant communities, describes a vast range of systems. Neurons in the brain mostly connect within one cluster, but occasionally their axons extend far outside, to an entirely different cluster. Proteins in a cell mostly interact within one functional group, but occasionally they connect with receptors far removed. Sites on the internet mostly connect within one tight group (celebrity news sites link to other celebrity news sites; biology sites link to other biology sites), but occasionally a site will connect far outside its cluster (TMZ will link to a study on neuroscience). The Kevin Bacon game had shown that there are surprisingly few steps between any two nodes (actors) in these kinds of networks. So Watts and Strogatz called a system with mostly local connections but occasional distant ties a "small-world network."

Coming back to the crickets, Watts wondered whether percolation had ever been studied on a small-world network. He assumed a question that basic must have been solved already, so he went to the library to look up the answer. No one had asked it. He asked Strogatz, who knew that no one had studied the question, and realized they were onto something bigger than insect musicology.

Whether a computer virus spreads widely across the internet or disappears quickly; whether a tiny neuronal misfiring is harmless or erupts into a seizure engulfing your brain; whether an idea spreads explosively throughout a population or fades away quickly—all are governed by similar dynamics: percolation on a small-world network.

Watts and Strogatz's paper was published in June 1998. As of mid-2018, it has been cited 16,505 times. Of the 1.8 million papers published in scientific journals on the topic of networks, their small-world paper ranks #1. It has been cited more than Einstein's papers on relativity, Dirac's paper on the positron, or any paper in history published on "fundamental" physics.

* * *

Earlier we heard Sherlock Holmes present the axiom of emergence: while individuals remain puzzles, man in the aggregate "becomes a mathematical certainty." Holmes was in pursuit of a burglar in that scene from *The Sign of the Four*, calculating the odds, explaining to Dr. Watson his theory of the criminal class.

A century after Arthur Conan Doyle wrote those words, a physicist from Oxford University began pursuing terrorists. He applied the principle of percolating clusters of fires in forests, poised to erupt, to percolating clusters of small-world networks, poised to erupt.

His strategy for tracking terrorists was based on a mathematical certainty.

THE POWER IN THE TAIL

You can only eat so many falafels in a week. So in the late 1980s, when he was a graduate student studying physics at Harvard, Neil Johnson would occasionally abandon the falafel truck on the street outside the Jefferson physics lab and eat lunch at the law school cafeteria next door. There he met Elvira Restrepo, a law student from Colombia. They married soon after and lived briefly in Bogotá, until Johnson was appointed a professor at Oxford in 1992 and they settled in England.

Johnson's work on guerrilla warfare and terrorism was inspired by a strange observation. "We would drop in to Colombia to visit family," Johnson told me, "and the news [from Colombia's decades-long civil war] would be something like: Three dead tonight. Eight dead tonight. Two dead tonight."

Johnson is a fair-haired Brit with an eager laugh and a populist's light touch with science. Imagine a young Tony Blair (when he was popular) explaining calculus, and you have Neil Johnson. But when he described the newscasts, the laugh disappeared. The reports brought back memories. "I grew up in London where it was: Here's the news from Northern Ireland. Two dead tonight. None dead tonight. Four dead tonight."

During his time at Oxford, Johnson had specialized in using the techniques of physics to find hidden patterns in what seemed liked random numbers. So when the second Iraq war began, in 2003, and daily death tolls once again made headlines, Johnson began to wonder: was there a pattern to those tragic daily numbers?

Johnson gained access to detailed data on casualties from Colombia's ongoing civil war. The casualties, he discovered, followed a pattern seen, but never explained, in stock markets.

Textbooks on the behavior of stock markets often begin, like the Bible, with a declaration of faith. In the beginning, there were efficient markets. Markets capture all information into their prices; deviations from efficient prices are random (often called "random walks"). Bad actors can spoil the show (insider trading; manipulation), but with good behavior and proper enforcement, markets will revert to pure, perfectly efficient form. Much of modern finance theory, including estimates of risk and the pricing of stock options, is based on this belief.

Real markets, however, don't seem to work this way. Price movements that should happen once a year instead happen daily. Stock exchanges in New York, London, Paris, and Tokyo all show the same pattern. The curve that measures the frequency of price movements is supposed to have a minuscule tail, which accounts for rare outliers. In the real world, those tails are not minuscule. When extreme outcomes happen much more frequently than you expect, the probability distribution develops what statisticians call a "fat tail."

Physicists love fat tails. Random systems with no hidden connections, like coin tosses, have thin tails. They're kind of boring. Fat tails signal interesting dynamics in a network. That might be a network of trees through which a fire spreads. Or it might be a network of people trading stocks, through which an idea spreads—in other words, a financial market. Physicists, including Johnson, had been studying the fat tails in financial markets for years, trying to make sense of them. Market crashes (Greenspan's "notably rare exceptions"), hedge fund collapses, and sudden bank defaults are often caused by, or at least associated with, fat tails.

In 2003, Johnson coauthored a textbook on the physics of finance—applying techniques of statistical physics to markets. The book contained an offbeat suggestion. Most researchers tried to solve the fat tail problem by studying the behavior of individual traders. Johnson, instead, looked at clusters. He asked what would happen if we assumed traders acted in cliques: small groups whose members all behave the same way, that is, they make the same buy or sell decisions. (The evidence for groupthink in markets, from tulip mania to the internet bubble, is strong.) The clusters need not be permanent. Just like cliques in high school, members come and go, trading cliques form and dissolve, they merge with other cliques or split into two. Imagine bringing a pot of water to a boil. Just before the boiling point, bubbles of gas appear. Those bubbles grow or collapse, merge with other bubbles or fragment, all while new bubbles are forming. Johnson proposed that trading cliques act like those percolating bubbles.

By building a model that was *simple, but not simplistic*—that is, it captured the essence of trading, without getting lost in the details—Johnson showed that his trading cliques model seemed to explain the fat tail distribution in financial markets pretty well. That fat tail took on a characteristic shape: a power law. There were 32 times fewer cliques of 40 people than cliques of 10. There were 32 times fewer cliques of 160 than cliques of 40. And so on. The number of cliques decreased with the size of the clique by an unusual power: 2.5.

The data on casualties from decades of civil war in Colombia showed a near-perfect power law as well. There were 32 times fewer attacks with 40 casualties than attacks with 10 casualties. There were 32 times fewer attacks with 160 casualties than attacks with 40 casualties. The number of recorded attacks decreased with casualty size by the same unusual power: 2.5.

The similarity between trading data and one set of guerrilla warfare data, from just one country, could be a coincidence. But it would be a strange coincidence. Such a neatly ordered power law is rare. So Johnson and his collaborators began looking at other conflicts. Remarkably, data from wars in Iraq and Afghanistan showed the *same* pattern: casualties from attacks followed the same power-law form, with the same 2.5 exponent. Over the next three years, they recruited help and data from a broader set of researchers around the world, eventually assembling a database of 54,679 violent events across nine wars (or "insurgent conflicts"): Senegal, Peru, Sierra Leone, Indonesia, Israel, and Northern Ireland, in addition to their original three—Iraq, Colombia, and Afghanistan. The pattern persisted: a power law with an exponent of 2.5.

Just as Johnson and his group were compiling their data, another group of researchers, based in Santa Fe, New Mexico, reported on casualties from global terror attacks, using the largest database of terror attacks available, with records of over 28,445 events in more than 5,000 cities across 187 countries. The events spanned four decades, from 1968 to 2006. Whether analyzing deaths alone, or injuries plus deaths, the data showed a surprisingly strong statistical pattern: a power law with an exponent of roughly 2.5.

The common pattern was a clue, but not definitive evidence of percolating clusters: groups that form and dissolve, merge or fragment, in an endless cycle. There are many possible explanations of power laws (although very few that naturally come with an exponent of 2.5). Johnson needed stronger evidence.

In forests, you gather evidence by taking aerial pictures and tracking the progress of fires over time. Fire clusters form and burn out, merge or fragment. Aerial photography, however, can't help you track cliques of people. And asking terrorists to fill out a questionnaire about their social habits (Please list any terror groups you joined or left recently!) did not seem like a winning research strategy. Johnson and his team were stuck with an intriguing but inconclusive hint.

Until 2014, when ISIS emerged, and Johnson decided to look online.

WHEN TERROR GOES VIRAL

Tracking interest in terror activity by individual users on social media—sympathetic posts or tweets in reaction to events, for example—has proven to be a poor predictor of future attacks. Johnson's data, however, pointed to analyzing *clusters* rather than individuals. So Johnson looked for signs of online clustering.

Johnson and his team quickly discovered that ISIS-interested followers were forming ad hoc groups on VKontakte, the largest Russian social network. They came together by linking to a common virtual page (the equivalent of a Facebook fan page for a brand or a business). Facebook immediately shuts down pro-ISIS pages. The Russian site, however, which had 350 million users at the time, does not. Because the groups stay open to attract new followers, Johnson and his team could track pro-ISIS pages closely. Followers used their common page to post real-time battle updates, teach practical survival skills (how to evade drone attacks), request funds

Sample content from an online terror cell

(for fighters who wanted to travel to Syria but couldn't afford it), and, of course, recruit ("This is a call to all brothers!").

These online groups—virtual terror cells—are not fixed hubs in the familiar sense of a bus station, for example, where people gather to take a bus. Everyone knows where the bus station is. It was there yesterday and it will be there tomorrow. A bus station doesn't suddenly materialize, grow, dissolve, merge with another bus station, or split into two smaller bus stations.

The online terror cells, on the other hand, do all of the above. Just like cliques in high school. Or traders in financial markets.

Terror cells in the offline world are extraordinarily difficult to identify and track. Johnson and his team soon realized that the virtual terror cells, by contrast, are easy to track. Simple computer algorithms can detect and record when new users link into a virtual cell, when followers unlink and leave, when cells merge, when cells split up, when cells are hunted by online agents and rapidly dissolve, when those followers reassemble into new cells, and so on.

From the early emergence of ISIS in 2014 through the end of 2015, Johnson's team collected minute-by-minute data on the online behavior of 108,086 individual followers linked to a total of 196 of these virtual terror cells. It may be the largest publicly available forensic data set ever assembled on terrorist behavior.

The figure below shows one snapshot of the network, where the individual followers are the smaller dots, and the pages they connect through, the virtual terror cells, are the larger dots.

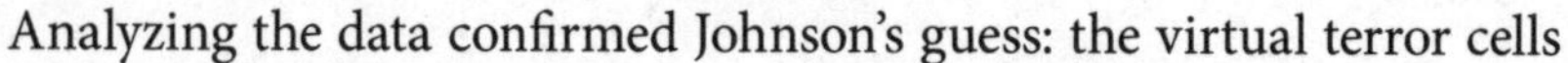

Analyzing the data confirmed Johnson's guess: the virtual terror cells behaved like percolating clusters. They grew, merged, split apart, or collapsed just like fires in a forest. In forests, the two control parameters are the density of trees and the likelihood that a fire will spread from tree to tree ("virality"), as shown in the forest fire phase diagram earlier in this chapter. Below the critical threshold, small fires die out. Above that threshold, they erupt into a wildfire.

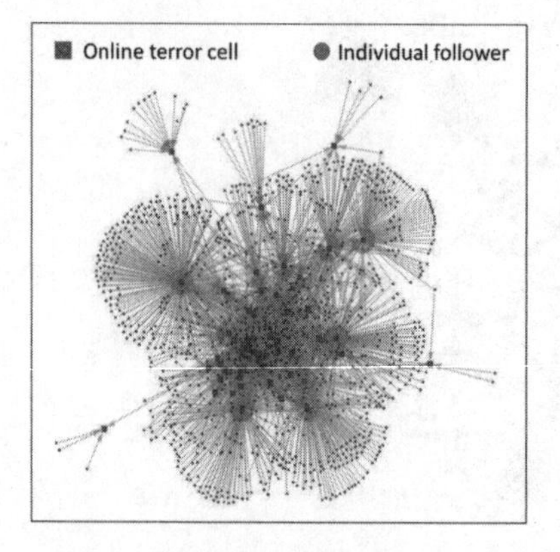

Map of an online terror network

Johnson's team identified similar

control parameters for the virtual terror cells on the Russian website. The number of clusters was like the density of trees. The rate at which one follower linking into a node inspires another follower to link into a node—the "infectability" of the cause—was the equivalent of the rate at which fire hops from tree to tree, the "virality."

Extrapolating from the model of fires in a forest, Johnson and his team could then predict when those control parameters would cross a critical threshold and the network would erupt. In other words, when an attack was imminent.

To test the theory, Johnson's team analyzed not only data from terror attacks but also, working with national authorities and using the same techniques, data from online groups for civil protests in Latin America. They found that signatures of incipient attacks and mass protests appeared weeks in advance. The figure below shows one measure of how the terror network grows exponentially before erupting, a signature that could predict the timing of an attack to within days.

Applying these percolation-style models to virtual terror cells has opened the door not only to new methods of detection and prediction but to new strategies.

First, the results suggest that rather than having to closely monitor

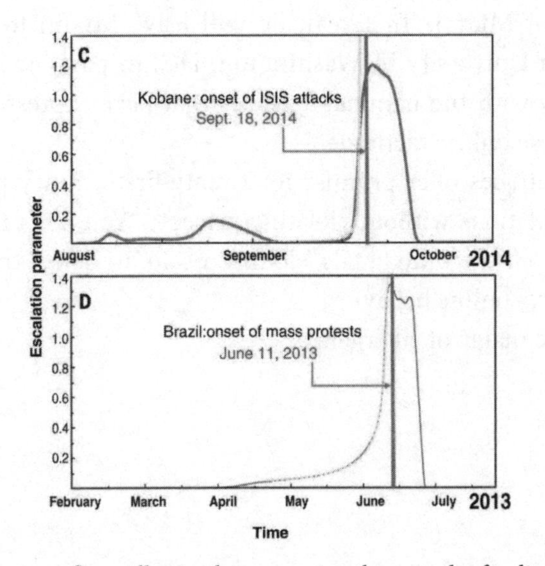

Predicting when conflict will erupt by measuring the growth of online terror cells

millions of individual online behaviors, focusing on the behavior of a small number of cells, which may number in the tens or hundreds, is a better use of time and resources.

Second, recently developed mathematical techniques can identify "superspreaders": clusters with the greatest influence. (Those are not always the ones with the greatest number of links.) The small-world networks found everywhere, described in the Watts-Strogatz paper, have an intriguing feature. They are *both* unusually robust and unusually fragile. They are robust against random attacks or random failures. Which is why random server outages, for example, have little effect on internet traffic. But they are especially vulnerable to attacks against the nodes with the greatest influence, as has been seen with attacks on the internet. Identifying and shutting down the online superspreaders is one strategy for fighting the spread of terror networks.

A third strategy is to increase the fragmentation rate—the rate at which clusters dissolve. The goal is to back a terror network away from the contagion transition, just as prescribed burns back a forest away from its contagion transition. (The authors writing on these topics are reluctant to discuss specifics.) Many more such strategies are being developed. And the techniques are being extended beyond ISIS to school shootings, bombings by nationalist groups, and other forms of violent conflict.

In 2007, Johnson left his job at Oxford for a faculty position at the University of Miami. This year, he will leave Miami to join George Washington University in Washington, DC, in part, he said, to work more closely with the national agencies that have expressed interest in applying these online methods.

The techniques offer promise for twenty-first-century policing: protecting populations without violating privacy. "You don't need to know anything about the individuals," Johnson said, to detect the patterns in their collective online behavior.

That's the magic of emergence.

§

THE MICROSCOPIC TUG-OF-WAR

Systems snap—liquids suddenly freeze, traffic suddenly jams, forests or terror networks suddenly erupt—when the tide turns in a microscopic battle. Two forces compete, and the victory flag changes sides.

The marble is drawn to the bottom of its well in the egg carton. But shaking the carton hard enough rocks the marble out of the well. That's binding energy vs. entropy.

A driver wants to cruise fast. But the driver brakes to avoid hitting the bumper in front of him. That's speed vs. safety.

Fires propagate from tree to tree, but they can exhaust their fuel, or rain can wet the trees. Violent causes can spread, but the ideas can get stale, or online agents can shut down virtual terror cells. Both are examples of increasing vs. decreasing virality.

Acting on a lone atom or individual, the forces cause only gradual change. But multiplied a thousandfold or millionfold, the change becomes the sudden snap of a system: a phase transition.

Now let's see how to apply these ideas to the behavior of teams, companies, or any kind of group with a mission.

7

Phase Transitions, II: The Magic Number 150

Why size matters

In the last chapter, we saw how a tug-of-war between two opposing forces can trigger a phase transition. As the temperature of water falls, molecules vibrate more slowly until they reach a critical temperature, at which point their binding energy exceeds their entropy and they crystallize into the rigid order of ice. That's the liquid-to-solid phase transition.

In this chapter, I will show you how something similar happens inside organizations. As a group grows, the balance of incentives shifts from encouraging individuals to focus on collective goals to encouraging a focus on careers and promotion. When the size of the group exceeds a critical threshold, career interests triumph. That's when teams will begin to dismiss loonshots and only franchise projects—the next movie sequel, the next statin, the next turn of the franchise wheel—will survive.

Even more important, we will see how to *control* that transition: how to change the magic number.

MORMONS, MURDER, AND MONKEYS

On the afternoon of June 27, 1844, a mob gathered outside a tiny jail in Carthage, Illinois. The pair of brothers inside had escaped angry mobs, or the law, a half dozen times across four states. This time their odds did not look good. Earlier that morning, the younger brother, Joseph Smith, had written to his wife, Emma, "I am very much resigned to my lot, knowing I am justified, and have done the best that could be done."

The jailer was friendly, so the brothers sent a visitor for a bottle of wine. "It has been reported by some," wrote a cellmate, John Taylor, years later, that sending for the wine "was taken as a sacrament. It was no such thing; our spirits were generally dull and heavy, and it was sent for to revive us."

The brothers asked Taylor for a song. He chose a spiritual hymn about a wandering man of grief, found in a prison, "condemned to meet a traitor's doom at morn."

"It was very much in accordance with our feelings at the time," Taylor wrote. The wandering man is revealed, at the end of the song, as the Savior.

A few minutes after Taylor finished, they heard shots fired, followed by pounding footsteps. Their cell door burst open; muskets blasted; Joseph fired a pistol that had been smuggled into the cell; Taylor tried to beat the attackers off with a cane. Minutes later the brothers were dead. Taylor, who was shot in the leg, had hidden under a bed, and was rescued.

Thirty-six years later, Taylor would become president of the religious order the two brothers had founded, called the Church of Jesus Christ of Latter-day Saints. The followers were initially called, derisively, Mormonites, and are today called, proudly, Mormons.

* * *

In the two decades since Joseph Smith's first visions in the early 1820s in a small farm town in upstate New York, the Mormon Church had grown to over 25,000 followers. Announcing a vision and organizing believers in the New England of this era was not uncommon. In Maine, the visions of Ellen White launched Seventh-day Adventism. In New York, visions of Revelation inspired followers of Jemima Wilkinson to build a town called Jerusalem. At the Harvard Divinity School, the poet Ralph Waldo Emerson (son of a minister) lectured that the true message of the living Jesus was that anyone could have spiritual visions and awaken others: "Cast behind you all conformity, and acquaint men at first hand with Deity."

Those other visionaries, however, stayed local. Smith's visions directed him west, to seek a New Jerusalem for his people. Everywhere Smith and his followers landed and built towns—Kirtland, Ohio; Jackson County, Missouri; Hancock County, Illinois—their otherness, as well as growing economic and political influence, threatened earlier settlers. The governor of Missouri had issued an executive order that Mormons must be "exterminated or expelled from the state." (The general of the Missouri militia that surrounded their town and confiscated their belongings explained that if they didn't leave immediately, they would be killed.) The town of Carthage adopted a similar resolution. Like another wandering tribe thousands of years earlier, this tribe found itself first shunned, then scapegoated.

In early 1844, frustrated with the inability of existing political parties to protect his people, Smith declared himself an independent candidate for president of the United States. The candidacy raised the level of his threat to civic leaders. A dozen sworn testimonies confirm a plan to kill Smith at the Carthage jail.

One year after the jail shooting, or assassination, a trial of six anti-Mormon leaders and members of a local militia, identified by eyewitnesses as the killers, resulted in full acquittals. The governor of Illinois feared that armed Mormon retaliation could escalate into a civil war. He urged Smith's successor, Brigham Young, and his followers to leave the state. Soon the urging became more insistent: leave or be forcibly expelled. Young agreed to go.

Young now faced a serious organizational challenge. How do you plan an exodus? How should you move thousands of families and their horses, mules, oxen, cows, sheep, pigs, chickens, dogs, cats, geese, and goats, all while searching for a permanent home? Young stewed on the problem, debated with his advisors, and finally, on January 14, 1847, announced that the Lord had spoken to him. The Church should divide into small companies, each led by a single captain, and head west.

Young led the first company of 149 people. Their thousand-mile trek past the Rocky Mountains ended when he saw an empty flat of land surrounded by mountains and streams and announced, "This is the right place" (now Salt Lake City). Fourteen more companies followed over the next twelve months.

The average size of those companies: 150.

* * *

A century later, Robin Dunbar, a researcher in the Department of Anthropology at University College London who specialized in the social habits of gelada baboons, published an unusual article.

By way of background, Dunbar is not your average primatologist. This is Dunbar on the grooming habits of monkeys:

> To be groomed by a monkey is to experience primordial emotions: the initial frisson of uncertainty in an untested relationship, the gradual surrender to another's avid fingers flickering expertly across bare skin, the light pinching and picking and nibbling of flesh as hands of discovery move in surprise from one freckle to another newly discovered mole. The momentarily disconcerting pain of pinched skin gives way imperceptibly to a soothing sense of pleasure, creeping warmly outwards from the centre of attention.
>
> You begin to relax into the sheer intensity of the business, ceding deliciously to the ebb and flow of the neural signals that spin their fleeting way from periphery to brain, pitter-pattering their light drumming on the mind's consciousness somewhere in the deep cores of being.

The passage made me, briefly, want to be a monkey.

In his 1992 article, Dunbar listed measures of brain volume and average social group size for 38 species of lemurs, monkeys, and apes. He showed that if you plotted one measure of brain volume (size of neocortex) vs. social group size, the plot seemed to lie along a straight line—the bigger the brain, the larger the group.

So Dunbar proposed a novel idea: the size of a species' brain determines the optimal size of their social groups. Maintaining relationships, argued Dunbar, requires brain power. More relationships require more neurons. Extrapolating his straight line from primate brains to human brains, he found that the optimal human group size, if this hypothesis were true, would be an interesting number: 150.

Despite Dunbar's obvious gifts for writing about monkeys, the article attracted little attention. Then, in 2000, Malcolm Gladwell published *The Tipping Point*, a blockbuster that included a chapter titled "The Magic Number 150." The chapter summarized Dunbar's monkey brain vs. group size results, as well as Dunbar's observations that mean group sizes in some hunter-gatherer societies and the "smallest independent units" of professional armies cluster around this number. Gladwell added the interesting example of Gore Associates, the maker of Gore-Tex fabric, which limits how many people work together in one building. "We put a hundred and fifty parking spaces in the lot," the president, Bill Gore, said. "When people start parking on the grass, we know it's time to build a new plant."

The idea of a hard-wired cutoff at 150 human relationships, set by volume of the human brain, went viral. Dave Morin, an early employee at Facebook, consulted with Dunbar and created a new type of social network, Path, based on the idea of limiting everyone to 150 friends. Minerva, the recently launched elite online university, plans cohorts of 150 students, citing Dunbar. Popular business and sociology blogs continue to spread the idea of "Dunbar's Number."

All of which has generated, predictably, a backlash among scientists. Dunbar anticipated one objection in his original article: extrapolating a straight-line correlation well beyond the range of the original data set is scientifically questionable. Half the monkey species in Dunbar's sample weigh less than a small pumpkin. To scientists, extrapolating from pumpkin-sized monkeys to humans is like analyzing a Mini Cooper to predict the behavior of a fully loaded tanker truck. There's also no biology that supports the idea. The link between number of neurons and primate behavior is dubious, at best, just like the link between number of genes and behavior. Onions, for example, have five times the DNA that humans have. To biologists, analyzing the volume of midget monkey brains to explain human behavior is, um, bananas.

Anthropologists and sociologists objected to the specific number. Many noted that hunter-gatherer tribes and army units have organized effectively into groups in a broad range of sizes. In the business world, some teams and companies have succeeded in remaining innovative at much larger sizes.

Just because a *theory* might be a bit wacky, however, doesn't mean there isn't something to the *observation*. In physics language, you might have the right observation but the wrong theory. Dozens of theories of superconductivity, for example, came and went; the observation never changed. Well before Dunbar's theory and other social models, Bill Gore and Brigham Young limited groups to 150 people. We intuitively understand that something changes inside teams and companies as they cross a certain threshold in size. But the volume of our neocortex might have nothing to do with it.

Let's see how the science of emergence offers an alternative explanation—and more.

THE INVISIBLE AXE

Imagine you are a middle manager at Pfizer. You attend a committee meeting to evaluate a project, an early-stage new drug. Like every early-stage project, it has warts. Some important experiments haven't been done, or have been done poorly. The science isn't really in fashion. Keynote speakers at big conferences dismiss the entire field. But you like the idea. Something about it captures your imagination. What do you do?

Option #1. You could pound the table, make the case, and begin the long slog up the ladder of committees, making the same case, pounding the same table, at each meeting of each committee. You might be turned down. But suppose you win that battle, and the next few. You make it all the way to the top of the ladder and win the green light to go ahead. The next seven years will be spent struggling to survive the Three Deaths of the loonshot. Each time the project stumbles, as it inevitably will, the smiling back-patters, who wished you well at the beginning, will turn around and try to bury you and your project. They want your budget. And they want you out of their way.

The odds that this loonshot succeeds—the drug works and people want to use it—are roughly one in ten. An outstanding drug might achieve $500 million in annual sales within its first few years of launch. Which means that the success will move the needle for your 100,000-person, $50 billion company by 1 percent. If the project does succeed, even with that tiny percentage benefit, you can count on 99,999 other people rushing in to claim credit. If it fails, you can count on 99,999 people backing away, pointing at you. They will mention all those early-stage warts, which you ignored so recklessly. Your career will be tainted. You might be fired.

Option #2. You could amusingly belittle the loonshot project, highlighting its flaws, poking at the warts. You explain why the world is moving away from just that kind of idea, why no keynote speaker at a big conference would ever mention it. You advertise your wit, breadth of knowledge, and good judgment to everyone in the room. By fortunate coincidence, your summary of where the industry is headed happens to agree precisely with what your boss and maybe her boss believe. They laugh along and nod.

You propose, instead, a modest step along a favored research direction. It's a franchise project, a known quantity. It's easy to green-light all the way to the top. Everyone gets it. If you continue to play smart politics, and sound good in meetings—funny put-downs, wise summaries of industry trends—you might just be able to get your boss's job. Perhaps as soon as next year. Which would increase your salary by 30 percent, not to mention double your prestige and influence. And the boost in title and salary could help you land an even higher-paying job at another company, when you start looking around. Just in case anything goes wrong.

So which do you choose? Do you pour your energy and ambition into Option #1, the loonshot with the seven-year slog, 1 percent return, and high chance of failure? Or into Option #2, the franchise project and politics with decent odds for a 30 percent bump in pay next year?

Dismissing the loonshot in favor of the franchise project is the rational choice.

Now imagine you work at a small biotech company, or a small production shop if this were a film project, or any kind of startup in a field with long timelines and low success rates. Instead of 100,000 people, there might be 50. The annual revenue, instead of $50 billion, might be zero. Success of the loonshot would increase revenue not by a percentage point or two, but by an infinite (or very large) percentage.

If you own a piece of the business—hard equity—the financial rewards could be worth millions. And there would be other kinds of rewards, too—the recognition from your peers, friends, family. We can call that nonfinancial stake in the outcome *soft equity*. You gambled and you won. You made a difference. Just you and a small team of underdogs. You own that win forever.

In this case, sounding smart in meetings or trying to get your boss's job is irrelevant. What matters is the survival of the loonshot: coming together to rescue it from its Three Deaths, to carry it to glory.

Uniting to support the loonshot project is the rational choice.

As the small startup's size gradually increases, it will eventually reach a breakeven point where the two incentives, pulling in opposing directions, are equal. Above that size, a behavior appears across the organization that favors killing loonshots and supporting franchises. Let's call that behavior the Invisible Axe.

The sudden emergence of that Invisible Axe is a phase transition.

* * *

The example above is what physicists call a *Gedankenexperiment*, or thought experiment. It serves as a mental warm-up exercise to help us get a sense of the forces at work.

Part of the blessing, or curse, of a physics overeducation is the inability to let a good thought experiment lie without trying to raise the stakes. The equation below follows from a simple mathematical model on how the thought experiment above might play out in the real world.

THE TUG-OF-WAR

In the previous chapter, I described an important step in developing new insights into forest fires: creating a model that was *simple, but not simplistic*. Hemingway wrote that "the dignity of movement of an iceberg is due to only one-eighth of it being above water." He called it his Theory of Omission. The power of beautiful prose comes from what you leave out. In science, it's the same. The power of a beautiful model comes from what you choose to omit.

In that spirit, to understand what causes a phase transition inside organizations, we need a simple-model organization, one that captures just enough to illustrate the basic idea. Later, we can add bells and whistles to build fancier theories.

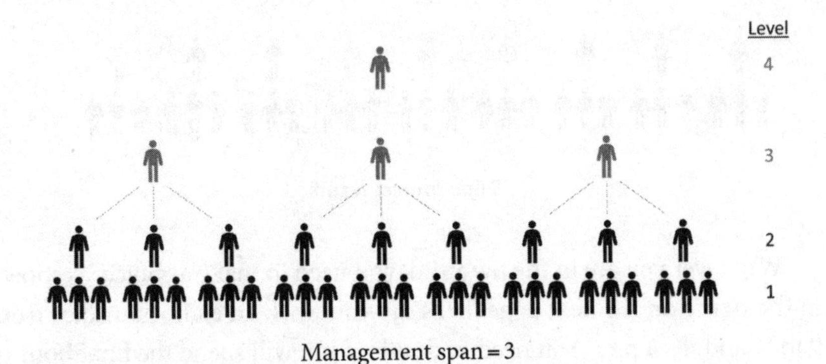

On the previous page is a picture of a simple-model organization. The super-senior manager at the top of the pyramid, who might be a vice president in a real company, has three direct reports. Each of the senior managers below that vice president has three reports, and so on, until the most junior level. The number of direct reports is called "management span." In US companies, the average management span has been between five and seven for many years, although recent studies have suggested the span has grown as high as ten. (I show a span of three so the picture can fit on the page.)

Now we need to connect the people to something customers pay for, sources of value. In professional service firms—law firms, consulting firms, investment banks, advertising agencies, design firms—clients typically pay per project. Clients of law firms and banks might pay for work on a transaction (a merger, or a public offering of stock). Clients of consulting firms might pay for a market research study. Clients of ad agencies, for a marketing campaign. And so on.

The work gets done by a project team. At a law firm, a handful of junior associates (level 1 below) may do the research; a senior associate (level 2 below) may supervise. Consulting firms, banks, ad agencies, design firms, and other service firms use similar models. At product firms, which make things rather than provide services, project teams might develop or sell a small product (a coffee machine) or a piece of a large product (the ignition system of a car).

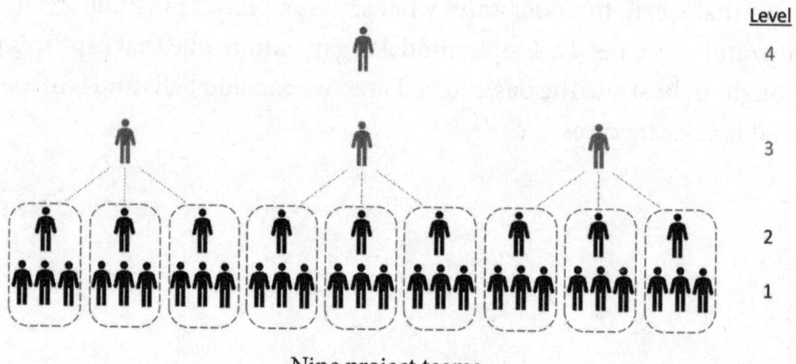

Nine project teams

Wherever you are in the pyramid, you need to make a choice, as shown in the figure on the next page. Let's say you work an eight-hour day, from 9 to 5, and it's 4 p.m. You need to decide if you will spend the final hour of

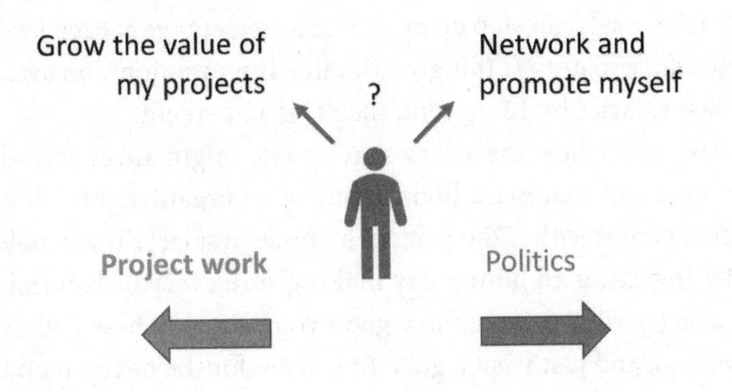

the day on (a) work that might increase the value of your projects (polishing up the client presentation; researching coffee machine designs), or (b) networking and promoting yourself within the company (currying favor with your boss, your boss's boss, or other influential managers).

Which will you choose? To capture the ideas in the thought experiment above quantitatively, let's look at your incentives.

The two main forms of incentives offered to employees go by many names and come in many flavors. Let's use the most common terms: salary and equity. Base salaries are examples of the first. By the second, I mean any form of payment that grants employees a stake in the success of their projects. Stock options are one example, but any "at-risk" pay connected to project outcomes qualifies—for example, restricted stock or profit sharing or bonuses.

"Hard" equity like stock options or bonuses, as mentioned earlier, is not the only kind of equity. People are motivated by more than just take-home pay: a passion for a higher purpose, the desire to be recognized and appreciated, the ambition to grow one's skills. The hard and the soft are not mutually exclusive. They are complementary. But focusing just on one and ignoring the other can be a mistake. Let's begin with the hard equity and come back to the softer, less quantifiable factors in the next chapter.

Design parameters *(G, S, E)*

In our simple-model organization, every employee has a base salary that depends on their level in the hierarchy (the levels on the right in the diagrams on the previous pages: 1, 2, 3, 4). To keep things simple, we'll assume that base salaries step up by the same percentage at each level. We'll call that *salary step-up G* (for growth rate). If promotions, on average, increase base salaries by 12 percent, then *G* is 12 percent.

It's easy to see how the salary step-up rate might affect your decision on how to spend that extra hour. Imagine an organization where every promotion comes with a 200 percent increase in salary. That would be incredible! Investing an hour a day making sure every influential person knows exactly who you are, how good your work is, how bad everyone else's work is, and just what a good fit you are for the next job up the ladder might be a very good use of your time.

If, however, promotions come with a 2 percent increase in pay, who cares? You might as well put your energy back into your project, where some extra effort could earn you a bigger bonus or increase the value of your stake in the company's success.

Management span, which we will write as *S*, is a second design parameter. To get a feel for management span, let's consider a company with roughly a thousand employees. If the span is very narrow, where each manager supervises three people ($S = 3$), there will be *five* layers between the CEO and the most junior level. If the span is much wider, ten direct reports on average ($S = 10$), then there will be only *two* intervening layers.

It's also easy to see how span affects your choice. Imagine an organization with an enormous span, more than one hundred direct reports (we will discuss one such example in the next chapter). Promotions happen so rarely that it's not worth spending any time politicking. With a span of two, however, you are constantly in competition with your peer. The career ladder may be always on your mind. There are so many layers that as soon as you get one promotion, you may immediately start thinking about the next one. The promotion obsession never stops.

Span also affects the balance of incentives in another way. In flatter organizations, all other things being equal, managers have a greater equity stake in the company's overall success. They own a greater piece of the pie. Which encourages focusing more on project outcomes and less on politics.

The third design parameter is *equity fraction*, which we will write as *E*. Equity ties your pay directly to the quality of your work. If you make a better coffee machine, the company may sell more coffee machines, so the value of your equity grows. If you write a better client presentation, the client may hire your company again and speak highly of the work to others. Your stake in that future income grows. The greater your equity fraction, the more likely you are to choose project work over politics.

Fitness parameters *(F)*

The three design parameters above are straightforward. Any company's HR (human resources) person should be able to quickly report an average *G*, *S*, and *E* across the company. But two more subtle parameters also shape your incentives.

Suppose you are wildly skilled at your job, or, at least, on the projects you have been assigned. You are so skilled that an extra hour per day invested in your projects might double, or triple, their value. You might design the greatest coffee machine in history, which would outsell every other machine in the industry. In which case, schmoozing your boss and networking with other influential managers inside the company is irrelevant, because your coffee machine triumph would speak for itself.

Suppose, on the other hand, you're not very skilled on the projects you have been assigned. One more hour doesn't matter—you'd design the same lousy coffee machine. In which case, you might as well invest that extra hour in politics and lobbying. The extra effort might help win you a promotion.

The greater your skill on the projects to which you have been assigned, which we can call *project–skill fit*, the more likely you are to choose project work. The lower your project–skill fit, the more likely you are to choose politics.

A final fitness control parameter is difficult to measure, but every employee feels it. Let's call it *return-on-politics*: how much politics matters in promotion decisions. Are promotions decided purely (or almost entirely) on merit? Or do lobbying, networking, and self-promoting make a big difference?

The answer will, of course, vary from manager to manager. Some managers are more susceptible to schmoozing and lobbying; some are less so. But just as height varies among individuals, yet every country has an average level, the importance of politics will vary among managers, but every company will have some average level. When we casually speak of one company being more "political" than another, this is what we mean.

Some companies actively seek to reduce the importance of politics in promotions by investing heavily in independent, exhaustive assessments (as we will see in the next chapter). Others are run like old-boys' clubs. A few senior managers sit around a table and decide who they will allow to join their club. The second kind of company is likely to be more political than the first.

What matters for our purposes, in our simple-model organization, is the ratio between *project–skill fit* and *return-on-politics*. We will write that ratio, which is a measure of overall organizational "fitness," as *F*. In high-fitness organizations, reward systems discourage politics and employees are well matched to their roles. As a result, they are eager to spend time on their projects—building the best coffee machine. In low-fitness organizations, politics strongly influence promotion decisions and employees are poorly matched to their assignments. As a result, they are inclined to spend time on politics.

Here's a more quantitative description: If an employee spends an extra 10 percent of their time on project work, by how much, on average, will he or she improve the expected value of those projects? By 1 percent, 10 percent, or 100 percent? (Expected value refers to the usual business-financial meaning: the probability-adjusted value of future income streams.) At companies where employees receive very little training or are assigned to projects carelessly with little consideration of how well the projects suit their skills, this measure of *project–skill fit* will be low. At companies that invest heavily in training, in recruiting the most talented people, and in carefully managing assignments, it will be high.

The *return-on-politics* can be defined in a similar way. If an employee spends an extra 10 percent of their time on lobbying and networking, by how much, on average, will he or she improve their likelihood of being promoted? By 1 percent, 10 percent, or 100 percent? The greater the likelihood, the higher the return-on-politics.

THE MAGIC NUMBER

When we examine the combined salary and equity incentives for individuals, we find that there is a critical size of organization, a magic number *M*, above which the balance flips from favoring project work to politics.

Below this threshold, incentives encourage employees to unite around making loonshots successful. Above this threshold, career considerations become more important and politics suddenly appears. Loonshots are more likely to be dismissed in favor of franchises. Although every person, individually, may enthusiastically believe in innovation, collectively the Invisible Axe emerges.

So what is this magic number? In appendix B, I show that

$$M \approx \frac{E\,S^2\,F}{G}$$

E	Equity fraction
S	Management span
F	Organizational fitness
G	Salary growth rate up the hierarchy

Let's see how this works. Since the equity fraction E is in the numerator, as E increases the magic number M gets larger. That means an increasingly larger group of people can work together, free of politics, in the loonshot phase. That makes sense, as we discussed above, since a greater equity stake encourages time on projects rather than on politics. The management span S is also in the numerator (at double power). Increasing management span reduces the number of layers, which reduces the importance of politics. It also increases an employee's stake in project outcomes. Both of which favor focusing on loonshots rather than career interests.

As the salary step-up G increases, however, the opposite happens. Large salary step-ups encourage politics, as employees compete to curry favor and win giant increases in salary. That decreases the maximum number of people, M, who can work together in the loonshot phase.

The last factor is organizational fitness F. If companies create review systems that guard against politics, invest in developing their employees' skills, and do a good job in matching employees with projects that allow their skills to shine—those companies will increase their likelihood of nurturing loonshots.

So what is this magic number M in the real world?

For management span, let's use the middle of the range mentioned earlier, six. And let's plug in a typical step-up in salary from promotions of around 12 percent. We will come back to typical values of E and F later, but for now let's consider an evenly balanced company, where equity and salary are equal fractions of pay (each 50 percent) and the skill and politics ratios are equal ($F = 1$).

Plugging in these numbers we find

$$M \approx \frac{50\% \times 36}{12\%} = 150$$

That's interesting.

I mentioned recent studies have reported larger management spans. In 2014, a survey of 248 companies conducted by Deloitte (one of the Big Four global accounting firms) reported an average span between nine and eleven. Along with that increased span and responsibility, of course, comes an increased growth in pay. Let's plug in a management span of ten and an aggressive, but not outrageous, average salary step-up of one-third (33 percent). We get

$$M \approx \frac{50\% \times 100}{33\%} = 150$$

Also very interesting.

Brigham Young, Bill Gore, Malcolm Gladwell, and Robin Dunbar may have been onto something. For typical real-world values of the control parameters there is, in fact, a sudden change in incentives around the magic number 150. At that size, the balance of forces in the tug-of-war changes, and the system suddenly snaps from favoring a focus on loonshots to a focus on careers.

The methods of Young et al., however—prayer, parking, and primatology—don't tell us what, if anything, we can do to *change* this magic number, so that larger groups can avoid this fate and succeed in nurturing loonshots. You will never see on a list of next steps or actionable takeaways something like: "increase the neocortex volume of our employees."

The science of emergence, on the other hand, does suggest some practical things we can do. We adjust control parameters to manage transitions. As mentioned earlier, we sprinkle salt on our sidewalks before a snowfall because salt lowers the freezing temperature of water. We want the snow to melt rather than harden into ice.

In the prior chapter we saw how phase diagrams capture the essence of the traffic flow and forest fire phase transitions and provide us with a guide for managing those transitions. So what does the phase diagram look like in this case—and what does it tell us?

Below the critical threshold, the magic number in the equation above, incentives encourage individuals to unite around loonshots. When group size crosses that magic number, incentives shift toward favoring a focus on careers: the politics of promotion (#1 → #2 in the diagram below). For typical group structures, that number may be roughly 150. But by adjusting the parameters of structure—equity fraction, fitness, management span, compensation growth rate—we can raise that magic number (which is why the dashed line is angled; if we couldn't adjust the number then the dashed line would be purely vertical). Another way to say the same thing: groups much larger than 150 people, which are stuck in the

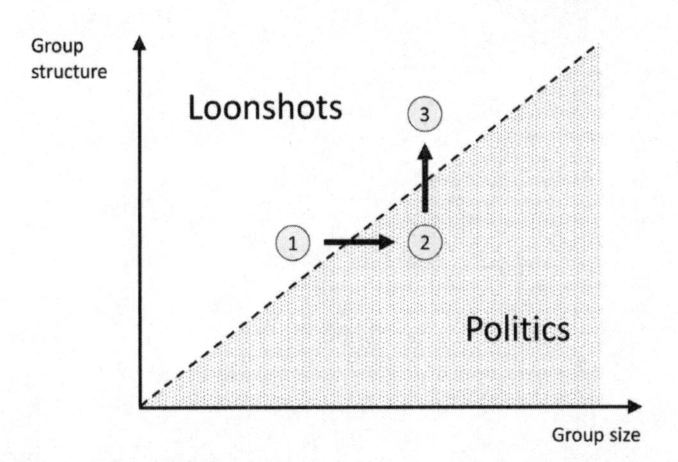

When group size exceeds the magic number (1 → 2), incentives shift from encouraging a focus on loonshots to encouraging a focus on careers: the politics of promotion. Adjusting structure can restore the focus on loonshots (2 → 3).

career politics phase (#2 in the diagram), can restore the focus on loonshots by adjusting their structure (#2 → #3).

As mentioned earlier, the ideas apply not only to small teams that want to remain entrepreneurial as they grow but also to large organizations that wish to create powerful teams for nurturing loonshots internally.

Let's see how a *two-million*-person organization created a loonshot group whose record of radical innovation may be unmatched by any other organization of the past half century.

8

The Fourth Rule

Raise the magic number

Since 1958, one two-hundred-person research group, deep inside a massive organization, has spun out the internet, GPS, carbon nanotubes, synthetic biology, pilotless aircraft (drones), mechanical elephants, the Siri assistant in iPhones, and more. Its alumni have led, or its management principles have inspired, many of the most legendary research organizations across the United States, including nearly every example mentioned in this book.

Those management principles are examples—some are extreme examples—of adjusting the parameters described in the previous chapter to raise the magic number. Those changes crank up the creative output of any loonshot team.

Before we dive in, it's worth keeping in mind that revving the creative engine to fire at higher speeds—nurturing more loonshots, with greater productivity and efficiency—means more ideas and more experiments, which also means, inevitably, more failed experiments. That's not the right choice for every team. If you're part of a group that assembles planes, for example, you don't want to launch ten planes and see which eight fall from the sky. The manufacture and assembly of planes belongs in the franchise group. The loonshot group is for developing the crazy new technologies that might go inside those planes.

And now to that two-hundred-person loonshot nursery. It all began with a steady beeping sound from a shiny metal ball in space.

WEB OF DARPA

Vannevar Bush created the Office of Scientific Research and Development, as described in chapter 1, to develop unproven technologies the military was unwilling to fund. Shortly after the end of the Second World War—to pick up where that chapter ended—the organization dissolved, without any successor. The national research agency that Bush envisioned in *Endless Frontier* was first vetoed by President Truman and then delayed due to political fights in Congress. After losing those battles, Bush wrote, "the stresses finally caught up with me and I proceeded to fold and go out of circulation." He would never fully reenter public service.

By 1950, the conflicts had been resolved and Congress established the backbone of today's extended system of national research labs. The National Science Foundation, the National Institutes of Health, and similar agencies support public-interest science: research on the spread of disease, water purification, earthquake prediction, and so on. The agencies also support research in areas that might be called "market failures": fields whose commercial futures are so uncertain or distant that no one company can afford to invest in them. Genetic engineering, for example, 50 years ago. Nuclear fusion, today.

The NSF and NIH are civilian agencies. No similar loonshot lab replaced Bush's OSRD within the military, even as the Cold War escalated. In 1949, the Soviets exploded their first nuclear weapon. In 1950, the Cold War turned hot in Korea. In 1952, Eisenhower campaigned on the Soviet menace and the strength of his military record ("If elected, I shall go to Korea"), and won. Two years later, in November 1955, the Soviets exploded their first hydrogen bomb. It was a hundred times more powerful than the bomb used on Hiroshima (in 1960, they exploded a bomb *three thousand* times more powerful than the Hiroshima weapon). One hydrogen bomb could wipe out the eastern seaboard of the United States.

Eisenhower and his military advisors quickly decided the Soviet show of force should be answered with a greater show of force—bulking up the US arsenal of missiles. Bigger, faster, more. Franchise projects. The lessons of Vannevar Bush and his national department of loonshots had faded.

And then, in October 1957, the Soviets launched a twenty-two-inch aluminum sphere, polished to make it more visible from earth, into orbit. Anyone with a pair of binoculars could look up and see the Soviet satellite, called *Sputnik 1*, as it passed over the United States. If they had a ham radio, they could hear its unnerving stream of beeps.

Launch of *Sputnik*, October 1957

For the first few days after *Sputnik*'s launch, Eisenhower shrugged. Its only purpose was for show. Eisenhower knew that US rockets could have placed satellites into orbit well before *Sputnik*, but he had no interest in expensive stunts. Eisenhower's opponents, however, sensed an opportunity. Senator Lyndon Johnson said that the Soviets "will be dropping bombs on us from space like kids dropping rocks onto cars from freeway overpasses." Another senator announced, "What is at stake is nothing less than our survival." The media sensed a story. The *New York Times* declared a triumph for communism; *Newsday* declared, "Russia Wins the Space Race"; the *Washington Post* declared that a secret report "portrays a United States in the gravest danger in its history." Interviewed on television, Edward

Teller, the father of the hydrogen bomb, offered that the launch was an even greater disaster than Pearl Harbor.

Eisenhower's secretary of defense had announced his retirement earlier that year. His replacement, Neil McElroy, sworn in just five days after *Sputnik*, was greeted by the media storm. Like Vannevar Bush, McElroy was an outsider. Unlike Bush, he had neither technical nor military experience. He began his career selling soap door-to-door for Procter & Gamble. Eventually he came up with the idea of shows that housewives could watch during the day, which P&G could use to deliver ads directly to their living rooms. "The problem of improving literary taste is one for the schools," he explained. "Soap operas sell lots of soap." His loonshot made billions for P&G. McElroy was also the father of brand management, the idea of small teams focused around individual brands (Ivory, Tide, Joy). Bright, young "brand managers" would run those teams like independent businesses. McElroy would soon replicate that idea in a very different context.

Eisenhower and McElroy understood the resistance to radical ideas inside large organizations. McElroy had overcome resistance to his crazy soap opera idea. During his long military career, the president had seen firsthand how interservice rivalries slowed progress, and he complained about it both publicly and privately. Which is why Eisenhower had selected McElroy for the job. He wanted an outsider with no prior ties to the military or federal government, free to shake up the system. The hype around *Sputnik* created an opportunity for some shaking.

On November 20, 1957, one month after *Sputnik*'s launch, McElroy proposed a new agency that would report directly to him and fund "far-out" research ideas. In other words, a group to develop unproven technologies that the military was unwilling to fund.

The military branch leaders hated the idea. They reacted just as their predecessors had to Bush's similar proposal at the start of World War II. The head of the Air Force, after receiving a draft charter for the new agency from McElroy, wrote back: "The Air Force appreciates that the proposals are suggestions." McElroy responded by explaining that these were not suggestions.

McElroy had helped to develop the nation's most successful consumer-product research lab. He understood that technologies buried in the lab that fail to transfer to the field, or labs that fail to respond quickly to feedback from the field, are useless. In other words, McElroy understood, and drafted into the charter for his new agency, the essence of the first two Bush-Vail rules: phase separation and dynamic equilibrium.

The resemblance to OSRD was no coincidence. Both Eisenhower and McElroy, neither of whom had any science background, were closely advised by scientists who had worked with Bush, including James Killian, the president of MIT, and Ernest Lawrence, a Nobel laureate, who helped launch both the microwave radar and Manhattan projects.

On February 7, 1958, McElroy's new organization officially began operations. He called it the Advanced Research Projects Agency.

Bush's OSRD had been reborn.

A GIANT NUCLEAR SUPPOSITORY

The stories of loonshots funded, or at least seriously considered, by ARPA (which was renamed DARPA in 1996, with "Defense" added) are legendary. One of my favorites, and I'm not sure what that says about me, was an idea presented as a giant nuclear suppository.

At the height of the Cold War nuclear fear in the early 1960s, a self-taught and widely respected physicist, known for his wild ideas, proposed using a giant particle beam to shoot down incoming Soviet missiles. Nicholas Christofilos had made his way from Athens, where he had worked as an elevator technician and taught himself physics on the side, to Livermore National Laboratory, which specializes in weapons research. In the forty years since Nikola Tesla made headlines with a death beam, technology had advanced to where Christofilos could discuss the idea seriously, or semiseriously, with a roomful of physicists. One objection they raised was the impractical cost of building giant tunnels to house the apparatus to generate the beam.

"There's a better way to do it," Christofilos said. Use nuclear weapons to build the tunnels.

"Think of it like a suppository," Christofilos explained, referring to his proposed new use for a nuclear weapon. "We would push it through the rock. As it goes into the rock, it melts the rock, it creates this perfect tube. You just have to keep on pushing it so it's hot enough so it melts the rock. As it goes, you just push it through."

Most physicists were left speechless by his ideas. The suppository project did not receive funding.

Many loony DARPA projects, of course, did get funded, and subsequently failed. A mechanical elephant to carry military equipment through the jungles of Vietnam. A superbomb made from the element hafnium, discovered, allegedly, by a physicist experimenting with a dental x-ray machine. A plan for achieving nuclear fusion through rapidly collapsing bubbles inside liquids (a modified cleaning fluid was used). A prediction market where investors could bet on the location of the next terrorist event, in order to tap into the "wisdom of crowds." (The project was scrapped for what might be called bad taste.)

Other DARPA loonshots have transformed industries or created new academic disciplines. The early computer network ARPANET evolved into the internet. A satellite-based geolocation system evolved first into military GPS, then the consumer GPS used in nearly every car and smartphone. A project to assist soldiers with software that could understand voice commands spun out Siri, now found in every iPhone. A worldwide system of seismic sensors, installed by DARPA to distinguish earthquakes from nuclear tests, made possible the first nuclear test ban treaty. (Other military branches had insisted that detecting nuclear explosions through tremors in the earth was impossible. A test ban treaty was pointless, they argued, and should not be pursued, because it could never be verified. They dismissed the small DARPA team pursuing the seismology idea as "a bunch of incompetents.") As an offshoot, the seismology project revived, and subsequently validated, the theory of plate tectonics. That theory permanently changed geology.

DARPA funded the creation of the first major computer graphics center. It chose the University of Utah. The group at Utah, described in chapter 5, was co-led by a former DARPA program manager, Ivan Sutherland. Sutherland supervised the computer graphics PhD thesis of Ed Catmull, the Pixar founder, who has said he was "profoundly influenced" by the DARPA model of nurturing creativity. DARPA funded another engineer, named Douglas Engelbart, who built the first computer mouse, the first bitmapped screens (early graphical interfaces), the first hypertext links, and demonstrated them in 1968 at what computer scientists now refer to as the "Mother of All Demos."

In 1970, much of Engelbart's team left and joined a newly created research group, led by another former DARPA program manager, Bob Taylor. That was Xerox PARC—the birth center of much of the early personal

computer industry. Taylor said he modeled that legendary research group on "the management principles developed at DARPA."

Former DARPA program managers or directors currently lead or have recently led research groups at Facebook, Google, Microsoft, IBM, Draper Laboratory, and MIT Lincoln Labs. This small group's management principles have spread throughout both private industry and public research in the US, forming an extended web of DARPA.

Let's see how those principles connect to the control parameters described in the previous chapter—the variables in our magic number equation. We'll see how raising the magic number can enhance radical innovation.

SIX DEGREES OF RED BALLOONS

Reduce the return-on-politics

In traditional structures, the career ladder is the ultimate carrot. Meet this goal, get a bigger office, a higher salary, more staff, and so on. That same career-ladder carrot, of course, encourages the weed of politics to spread.

DARPA is run like a loose collection of small startups, with no career ladder. A hundred or so program managers each lead one project or field of research. They are granted an extraordinary degree of autonomy and visibility, not unlike McElroy's brand managers. For example, in September 2009, Doug Wickert, an Air Force test pilot, rotated through DARPA as part of an exchange program with the service branches. Regina Dugan, a Caltech PhD in mechanical engineering, had recently been appointed as DARPA's director. At Wickert's first meeting with Dugan, she told him to come back with his team in two weeks and pitch her on an idea, "just like any other program manager."

At the time, Dugan and others at DARPA were looking for a way to celebrate the fortieth anniversary of the internet, recognized in computer circles as the 1969 launch of ARPANET. (The remote network went live on October 29, 1969, when Charley Kline's computer at the University of California in Los Angeles communicated with a computer at the Stanford Research Institute in Menlo Park, California. Kline typed the "l" and "o" of "login" and then the computers crashed.)

Wickert's idea was to test the power of the internet to unite people

around the country in solving a time-critical problem. So he floated an idea for a novel challenge—balloons. Red balloons. DARPA would place ten red weather balloons in ten undisclosed public parks across the country and see how quickly they could be found. "It was kind of a joke at first," Wickert told me, "but then it morphed."

One month later, on October 29, 2009—the internet's anniversary—Dugan announced DARPA's Red Balloon Challenge. The balloons would be placed in the parks, tethered to trees or benches, on the morning of Saturday, December 5. The first team to find all ten balloons would win a $40,000 prize. The short notice—just 37 days between announcement and competition—was intentional, to give teams limited time to prepare, just as similar teams might face in a real crisis.

Four weeks before the start, Wickert and his team tested inflating all ten balloons, each eight feet wide, in a field behind DARPA's office. Not everyone immediately got the hang of it. Wickert described with some glee the experience of "being an Air Force guy showing a Navy guy how to tie a knot."

Peter Lee, a former computer science professor who had just been hired to lead DARPA's efforts in new computer technologies, came down to the field and looked over the balloons. DARPA's oval logo had been painted on each one. "This is awesome," he announced. "The logo looks like a giant Eye of Ra!" Lee told Dugan, who realized that the sudden appearance across the country of government balloons painted with what appeared to be a giant eye might not be such a good idea. New balloons were quickly ordered.

At 10 a.m. on Saturday, December 5, the balloons—minus the Eye of Ra—went up in parks across the country, from Florida to Oregon. DARPA planned to take them down at 5 p.m. and repeat the process for up to a week. But just 8 hours, 52 minutes, and 41 seconds later a team from MIT found all ten sites. Even more amazing: the team had heard about the challenge only four days earlier.

The team's solution had been to launch a network with a creative reward system. A total pot of $4,000 would be assigned to each balloon. If Susan spotted a red balloon and reported it on the MIT team website, she would win half the pot: $2,000. If Greg was the one who told Susan about the game, he would win half of the remaining pot: $1,000. If Karen was the one who told Greg about it, she would win half of the remainder: $500.

And so on, so that everyone in the reporting chain would get a cut (any money left over from the $4,000 when the chain ended would be donated to charity).

The brilliance of the system was that even people who never left their home and couldn't possibly spot a balloon had a reason to pitch in and help MIT win. The links were all tracked through the team's website (built in two days). Mostly, the balloon hunters reached out to local friends. But occasionally they reached out to connections far away. In other words, it was a small-world network: six degrees of red balloons. The team recruited 4,400 people in just 36 hours.

The second-place team, from Georgia Tech, had a three-week head start, and worked feverishly, "fueled by donuts, pizza, and adrenaline," to set up a Facebook page and a Google Voice number in addition to a search-rank-optimized website. Yet they only managed to recruit 1,400 people. The Georgia team had bet on altruism. If the team won, they had promised, they would donate all their proceeds to charity.

The Balloon Challenge turned out to have many surprising lessons that none of the participants who originally participated could have anticipated, in addition to the lesson—consistent with the theme of these two chapters—that incentives matter more than we think. Many of those lessons have been written up in prestigious scientific journals. They are important for understanding how to use modern networks to mobilize groups to solve urgent problems: how to find a missing child or soldier, for example, or marshal resources for disaster recovery. And they continue to be tested in new challenges. The US State Department, for example, ran one recently called the Tag Challenge. Teams had 12 hours to locate five "thieves" (actors), hidden across five cities in the US and Europe, who were identified only by a mug shot. (The same MIT team won.)

One of the reasons that crazy projects like the Red Balloon Challenge can succeed inside DARPA is that there is no career ladder. Project managers are hired for fixed terms, typically between two and four years (their employee badges are printed with an expiration date). DARPA's structure has eliminated the benefit of spending any time on politics, of trying to sound smart in meetings and put down your colleagues by highlighting the warts in their nutty loonshots so that you can curry favor and win promotions.

The DARPA team prepares for the Red Balloon Challenge

Use soft equity

DARPA replaces that traditional career-ladder carrot with another. Project managers are publicly identified and broadly known in their community. They are granted authority to choose their projects, negotiate contracts, manage timelines, and assign goals. The combination of visibility and autonomy creates a powerful motivating force: peer pressure.

When we think of equity stakes, we usually think of stock options or bonuses or something similar that ties employees financially to the success of their project. Those are tangible forms of equity. Recognition from peers is a form of *intangible* or *soft* equity. It can't be measured through stock price or cash flows. But it can be just as strong a motivator, or even stronger, as both a carrot and a stick.

"A soldier," Napoleon said, "will fight long and hard for a bit of colored ribbon." In the case of mid-level corporate soldiers given visibility and autonomy, the colored ribbon is recognition from respected peers. Imagine a computer graphics pioneer called to the podium at a graphics conference, presented a trophy, basking in the admiration of his colleagues.

That intangible can quickly become tangible. If your job is to partner with external peers to develop new ideas, and you are recognized by those peers as a strong manager, scientists or inventors or other creatives will want to work with you rather than your competitors. They will be more inclined to bring their next great loonshot to you rather than your competitor. That could make a very big difference to your career. And, of course, being well known and well regarded among peers creates the obvious benefit of future job offers.

Partnerships reduce the return-on-politics in still another way. External peers are more likely to be impartial judges, not susceptible to politicking. That impartial view, of both successes and failures, is crucial for a strong system mindset, as discussed in chapter 5. Did a program fail because the underlying technology just didn't work (hypothesis failure), or because the person running the project botched it up (operational failure)? Did a program succeed because the person did a terrific job, or did he make a bunch of critical mistakes and get lucky despite those mistakes? Fans watching a baseball game recognize the difference between a player who scores off a beautiful hit and a player who scores off a sure out that a fielder happened to let dribble through his legs.

DARPA's principles—elevated autonomy and visibility; a focus on the best external rather than internal ideas—won't apply the same way to every company (most companies are not faced with problems that might be solved by a giant nuclear suppository). But every organization can find opportunities to increase autonomy, visibility, and soft equity.

One example is the growing practice of open innovation. In open innovation, companies jointly develop new ideas, technologies, or markets with customers (usually early adopters or superfans) or business partners (for example, suppliers and comarketers). The Red Balloon Challenge is an example of one organization recruiting the best minds across the nation to help it think through an important problem in network theory: how to rapidly mobilize groups.

The practice is common in the tech world. Software companies routinely share unfinished products with tight-knit developer communities to quickly generate feedback. Biotech companies frequently work closely with university scientists (and, in a growing trend, patient groups) to develop products. Recently, the idea has spread beyond tech. Open innovation helped the Coors Brewing Company develop a cold-activated beer can—the mountain logo printed on the can changes from white to blue when the beer reaches the ideal drinking temperature. (According to Coors, that's 43 to 50 degrees Fahrenheit.) The practice helped Kraft Foods develop melt-resistant chocolate. Parents can thank open innovation for summers free of sticky chocolate goo.

Open innovation comes with a double bonus. Companies gain access to fresh ideas, often from exactly the kind of enthusiasts they want to engage, like the donut-and-adrenaline-fueled kids on the Georgia Tech Red Balloon team or the brilliant improvisers on the MIT team. At the same time, they also improve soft equity—peer recognition. Those advantages need to be weighed against the desire to protect competitive secrets.

As research has become increasingly fast-paced, many companies have decided that the double bonus of open innovation, especially the long-term gain of a more nimble organization, now outweighs those of the closed, more secretive model.

* * *

If DARPA performed radical surgery on traditional organizational design, McKinsey & Company's surgery has been more limited, but still effective. The company is something like a halfway house for academics taking their first steps off campus (I worked there for three years, shortly after leaving physics and before starting in biotech). It has 27,000 employees, over $10 billion in revenue, and has ruled the world of management consulting for decades, even as that world has rapidly changed.

The career ladder is a powerful motivator at McKinsey, as at most firms. But at most firms, local offices or functional practices decide on promotions. A California office, for example, will decide on California candidates; an auto industry group will decide on one of their own. At McKinsey, for important promotion decisions, a senior partner, chosen for his or her limited overlap with a candidate's office and practice, is brought in to conduct an independent evaluation. The distance reduces the influence of local politics. If Tom in San Francisco, for example, is up for partner, Marianne from Brussels may be recruited to interview as many as two dozen colleagues and clients about Tom's performance. The process may take as long as three months. The investigation is exhaustive: a partner once explained to me that your evaluator ends up knowing your strengths and weaknesses better than your mother. The lost time is expensive. It takes away from client work, which generates income.

But the short-term costs are a long-term investment in strength of organization. Leaders who order their employees to be more innovative without first investing in organizational fitness are like casual joggers who order their bodies to run a marathon. It won't happen, and the experience is likely to cause a great deal of pain. There is a time-consuming process for getting your body ready for a marathon. If you invest in the process, in gradually building fitness, even if you begin well below average, you will cross the finish line with a smile. (I trained with a group for a long-distance triathlon. The coach used that phrase all the time. We all finished. What he didn't explain was that the smile would be one part joy, one part masochistic agony. Which also sounds about right for a large organization that radically transforms.)

The DARPA model is extreme: reduce career politics by eliminating careers. McKinsey approaches the same goal in a less extreme way. It retains careers but invests heavily in reducing the subjectivity of promotion decisions.

A TOOTHPASTE PROBLEM

Increase project-skill fit

It's no surprise that the ability to innovate well is connected to employee skill. More interesting is *how* it matters. Project–skill fit measures employee skills against *currently assigned projects*. If an employee spends a bit more time on a project, does he increase its value not at all, by a modest amount, or by a great deal? If this value is low across your team or your company, one obvious possibility is that you have weak employees. If you have a kitchen appliance company and no one can design a good coffee machine, you may be doing a poor job at attracting talent, or a poor job in training your people, or both.

Another possibility, however, is that skill levels are fine, but your company is doing a poor job at *matching* employees and projects.

In my first or second year working as a consultant at McKinsey, I was assigned to a four-person project working with a consumer-goods company. It sold products you can find in every supermarket: soap, toothpaste, skincare, and so on. My previous work experience was with physicists, software engineers, and traders at investment banks. Personal hygiene products never ranked high on my curiosity list. And I didn't know the first thing about marketing. Some project managers take an interest in showing a new person new fields and new skills. Mine didn't. The project was a disaster. I added little value, felt terrible, and considered leaving.

At many companies, if you do a truly bad job on a project and get a bad report from a supervisor, you're fired. At others, just before firing it might occur to someone to intervene and see if the employee deserves another chance in a different role. McKinsey, however, dedicates a full-time team to managing project–skill fit. The person in charge scans for bad fits and steps in to rescue a bad situation, as he did in my case. He pulled me off that project and away from that manager and placed me where my skills were better suited. My back straightened, and I did fine the rest of my time at the company.

My toothpaste project was an example of an undermatch: skills or experience not up to the task. But poor project–skill fit can also result from an *overmatch*: skills so far above project needs that the employee has maxed out what he or she can contribute. Imagine assigning the coffee machine project to a young Frank Lloyd Wright. He'll do a good job, of course, but after a few hours he'll be bored. Employees who are not stretched by their assigned projects have little to gain from spending more time on them. Let's come back to the graphic in the last chapter of the stick figure choosing whether to spend the last hour of his day on project work or politics. If there's nothing more young Frank Lloyd can do to improve the value of his project, then he might as well spend the time explaining to his superiors how well suited he is for a big promotion and highlighting the feeble efforts of his competitor down the hall. "Overmatched skills reduce project–skill fit" is the mathematical expression of "Idle hands are the devil's workshop."

The goal is a string that is neither too taut nor too slack: employees stretched, on average, neither too much nor too little by their roles. Dedicating a full-time person to keep that string tight is expensive. As with the efforts to reduce politics, mentioned above, the expense is a long-term investment in fitness: it's putting in the weekly runs to achieve marathon performance. But the investment also comes with an immediate bonus: it helps attract talent.

Suppose you are fresh out of school choosing between job offers. Company A offers a typical package: work for this person or that group at such-and-such salary. Company B offers a similar package, but also a person or team dedicated to finding the right project for you, separate from any boss or other politics. Staff dedicated to making sure you are well matched to your new role reduces your career risks—the risk that you fail because of a bad fit and are fired by a disappointed or vindictive boss (as I might have been). It increases the chances that you will find something that excites you, which will help your career long-term. Even for the same salary and opportunity, you'd rather join Company B.

The importance of project–skill fit also changes how we should think about training. Managers usually invest in training employees with the end goal of better products or higher sales. Send a coffee machine designer to a workshop on product design and you will get better coffee machines. Send a sales manager to a marketing workshop and your sales may improve. But training employees has another benefit. A designer who has learned new techniques wants to practice them. A marketer with new skills wants to try them out. Training encourages spending time on projects, which reduces time spent on lobbying and networking. In other words, it improves organizational fitness.

The same principle applies all the way up the leadership ladder. Leaders well coached on group dynamics are likely to spend more time with their teams. It's fun working with high-performing teams who appreciate you. It's less fun to spend time with dysfunctional teams who hate your guts.

A SHREDDING PROBLEM

Fix the middle

In the previous section we saw how DARPA uses *soft* equity: nonfinancial stakes in project success, like peer recognition. Most large or midsized companies not only rarely tap into the power of soft equity but they do a bad job of using ordinary (hard) equity: stock options or bonuses. Large companies, for example, often use a steep equity grant curve: they award large stock options or cash bonuses at the highest levels (as much as 100 percent of base salary), and tiny amounts at junior and mid-levels (below 10 percent). That creates exactly the wrong incentive for the most vulnerable part of the organization—the dangerous middle.

At the lower levels of an organization, where one person oversees one product or service without depending on many others, evaluations are not too difficult. The coffee machine came out great or it didn't. The gaming app attracted users or it didn't. Clients loved the presentation or they hated it.

At the most senior levels of an organization, the CEO and board can keep an eye on internal battles, and they can intervene directly, as needed, to separate personal agendas from collective interests. A CEO and board span the entire organization and have the least to gain from turf wars.

It's the dangerous middle between these two levels that carries the greatest risk in the battle between politics and loonshots. Evaluations are more complex than at the lowest levels: there's not just one coffee machine, but many products or services that may depend on dozens of external and internal factors, only some of which are under a manager's control. And those same managers are far from the watchful eyes of a CEO or board, so the small fires of political agendas quietly smolder with no one to extinguish them. A wants B's budget; B wants C out of the way; D wants A's headcount; and so on. Steep equity grant curves—big bonuses at the highest levels, tiny ones at the lowest levels—just raise the stakes of those battles. The big bonuses are just one or two steps up the ladder for middle managers like A, B, C, and D—so close they can taste them. The steep curve creates a middle-manager version of *Survivor*: a giant jackpot for those who succeed in crushing their colleagues and staying alive.

If the prize for promotion were not as rich—if project success earned you the jackpot of your dreams, and promotion earned you no more than a used tissue—then the battles would not be as fierce. People would spend a lot more time creating great products or nurturing loonshots, and a lot less time stabbing each other in the back. Tilting the rewards more toward projects and away from promotion means celebrating results, not rank. Examples of celebrating rank include not just big increases in base salary, but any kind of special privilege: parking spots, a special cafeteria, trips to Hawaii for "executive workshops," and so on.

In the language of the prior chapter, celebrating results not rank translates to increasing the equity fraction E and reducing the salary step-up G. Both of which *raise* the magic number. In other words, they will make large groups more likely to innovate well. Recent academic studies have come to a similar conclusion. One group noted that "increased [wage] dispersion is associated with lower productivity, less cooperation, and increased turnover." Translation: a big *G* is a bad thing.

Shifting rewards more toward projects and less toward promotion—just like the changes required to reduce the return on politics mentioned earlier—is difficult. It demands a lot from managers. Writing a bonus check to everyone in a group for 10 percent of their base salary on a good year and nothing on a bad year is easy. A system with larger stakes and greater variability—one person earns 60 percent for their triumphant coffee machine, another earns zero for a flop—is much harder. Easily measured, easily understood goals need to be carefully designed and agreed upon. Performance needs to be assessed fairly to avoid violent end-of-year arguments. Difficult messages need to be delivered with actionable suggestions, so an employee sees a clear path to greater rewards in the future.

But the *most* difficult job in redesigning incentives may be the business-world equivalent of the Hippocratic Oath: first do no harm. It is surprisingly easy to unintentionally create perverse incentives.

Here's an example from a slightly different context. When the Dead Sea Scrolls were first discovered by Bedouin shepherds in a desert cave near the Dead Sea in modern-day Israel, archaeologists offered to pay the shepherds for each new scrap they found. That encouraged the shepherds to rip any scrolls they found into tiny scraps. The archaeologists had the right idea in theory but didn't think through the perverse incentive in practice.

Shredding the Dead Sea Scrolls

The same thing happens in the business world all the time. Pay contractors by the hour, and problems may multiply. Reward sales, and profits may disappear (customers can be bought). Reward the number of products launched, or the number of drugs that enter clinical trials, and recalls and failed trials may balloon. It *sounds* like a good idea to put big bonuses at the top levels and tiny ones at the lower levels. But it turns the vulnerable middle into a scene from *Lord of the Flies*.

Examining the unintended effects of well-intentioned goals has not received much study. One exception is a recent article, "Goals Gone Wild," which traces a handful of famous business disasters to poorly constructed goals. In the 1960s, for example, the Ford Motor Company was desperate to compete with smaller, cheaper cars from Japan. So the CEO announced an exciting stretch goal: the company would produce a new car that would cost less than $2,000 and weigh less than 2,000 pounds—the Ford Pinto. The goal and tight deadline, unfortunately, did not leave much time for safety checks. The fuel tank was placed just behind the rear axle with only 10 inches of crush space. The design flaw, as lawsuits later showed, led to a less-than-desirable new feature: on impact, the car could blow up.

There's no such thing as a perfect incentive system, but it's easy to stumble into a terrible system, as the Dead Sea Scroll archaeologists discovered.

Even more common is a useless system, in which rewards are handed out that do nothing. I'm still amazed by how often large companies compensate junior or mid-level employees on company earnings. If your project can move earnings by no more than a tiny fraction of a percent, how does a company-earnings bonus motivate you? You might as well put your energy into twiddling your thumbs and fooling your boss into thinking you are indispensable while enjoying the free ride if earnings go up. (Economists call a similar issue in the use of public goods the "free-rider problem.")

Bring a gun to a knife fight

Money spent on company-earnings bonuses would be much better spent on the people and processes needed to help managers think through the subtleties of incentives. Larger HR groups often have a compensation specialist. But those roles tend to be filled by rubber-stampers who apply cut-and-paste formulas.

Companies routinely appoint chief information officers, well-regarded technology experts, to create state-of-the-art computer networks. Imagine appointing a chief incentives officer, well trained in the subtleties of aligning value, who is solely focused on achieving a state-of-the-art incentive system. How much might politics decrease and creativity improve if rewards for teams and individuals were closely and skillfully matched to genuine measures of achievement?

Rewarding one person for designing one coffee machine is a simple example. Somewhere between simple one-person rewards and wasteful free-rider rewards given to everyone lies a valuable and critical sweet spot: rewarding teams for collective outcomes. Designing a team reward is tricky. It requires carefully thinking through both the benefits and possible perverse incentives of many different choices. The analysis goes beyond the normal experience of a rubber-stamper payroll person. In other words, it requires a strategic chief incentives officer.

A good incentives officer can also save money. He or she can identify wasteful bonuses (for example, the free-rider earnings bonus mentioned above) and tap into the power of nonfinancial rewards: peer recognition, reduced commute times, choice of assignments, freedom to work on a passion project, and so on. Which is another reason to think of the position as strategic: A chief revenue officer (head of sales) seeks the highest sales for a given sales budget. A good incentives officer will also seek the maximum return from a limited resource: the most motivated teams for a given compensation budget.

The science of understanding how people respond to subtle changes in environment has grown rapidly over the past decade (see the postscript at the end of the chapter). Companies with outstanding chief incentives officers—experts who understand the complex psychology of cognitive biases, are skilled in using both tangible and intangible equity, and can spot perverse incentives—are likely to do a better job than their competitors in attracting, retaining, and motivating great people. In other words, they will create a strategic advantage.

Many organizations are too small to hire a full-time person for this kind of role. When my company first started, we couldn't afford a full-time chief financial officer or technology officer, so, like many other small companies, we hired specialists part-time. Just as with CFOs or CTOs, small organizations can engage a specialist in the subtleties of incentives part-time. In your battle with competitors for talent and loonshots, incentives are a weapon. If your competitors are all using knives, maybe you want to get yourself a gun.

For the last word on why compensation is not all about cash, even at the highest levels, and why understanding the subtleties of incentives is important, we have the CEO of a successful European company responding to scholars conducting a compensation survey. He explained why, for him, certain intangibles are more important than cash:

> I'd rather be worth 100 million euros, have fun now, and enjoy people's respect when I am the senile chairman of my firm than be worth a billion and get paid fat dividends by a little **** with a Harvard MBA, who runs my firm and lectures me at board meetings.

Fine-tune the spans

Although discussions of cognitive biases are still uncommon inside compensation departments, people have been talking about management span (the number of direct reports per manager) for decades. The problem with most of the literature on the right management span for companies is the same as the problem with the question "What's the right temperature for tea?" The answer to the tea question, averaged over a large sample of people, might be room temperature. It's the wrong question and a useless answer. Half like hot tea, half like iced tea.

The answer to the management span question is a similar split. Wider spans (15 or more direct reports per manager) encourage looser controls, greater independence, and more trial-and-error experiments. Which also leads to more failed experiments. Narrower spans (five or fewer per manager) allow tighter controls, more redundancy checks, and precise metrics. Which leads to fewer failures. There's no right answer averaged across a company: we tailor the tools to the phase. When we assemble planes we want tight controls and narrow spans. When we invent futuristic technologies for those planes, we want more experiments and wider spans.

Loose controls is what Bill Coughran means when he says he would lead teams by "keeping the reins in enough so that we didn't degenerate into chaos." Coughran led computer research at Bell Labs for twenty years, then moved out to the West Coast and joined a small startup. Two years later, the Google founders recruited him to lead the company's engineering group. They had just removed nearly all the managers in the group, but a few months later, Coughran said, "they realized that maybe that wasn't the best decision in the whole universe. . . . So I was one of the people who got hired as adult supervision."

The engineering group, which eventually included over five thousand people, was responsible for data storage. When Coughran joined, Google was backing up the internet daily. Shortly after, it added billions of emails (Gmail, 2004) and videos (YouTube, 2006). Traditional models for storing data wouldn't work; Coughran needed radical solutions. He organized his team so that over a hundred engineers reported directly to him—at one point, the total reached 180. Each of the engineering directors in his group managed 30 or so people. The spans were wide and the controls were loose. He was encouraging experiments and nurturing loonshots. His teams succeeded: they developed the radical storage solutions Google uses to store billions of emails and videos. Some groups had competed, testing different solutions, but eventually the engineers had come together to support each other when those loonshots inevitably stumbled.

Which takes us to another reason a wide management span helps nurture loonshots: it encourages constructive feedback from peers. At Xerox PARC, for example, the entire computer research lab, between 40 to 50 people, reported directly to Bob Taylor. The structure, one engineer said, "provided a continuous form of peer review. Projects which were exciting and challenging obtained more than financial or administrative support; they received help and participation from other [lab] researchers. As a result, quality work flourished, less interesting work tended to wither." More layers "would have promoted organizational distractions, tempting researchers to worry more about titles and status than problem solving."

Taylor and Coughran understood about engineers what Catmull understood about film directors: creative talent responds best to feedback from other creative talent. Peers, rather than authority. Catmull designed a system for a group of peer film directors to regularly coalesce around a project and give its director advice—honest feedback from colleagues rather than marching orders from marketers or producers. Creatives are suspicious of those outside their faith. It's similar in drug discovery: biologists and chemists respond best to criticism from their own kind, much less well to suggestions from MBAs.

A wide span encourages creatives, whether film directors or software designers or chemists, to come together and help a colleague solve a problem. A span of two, on the other hand, encourages sabotaging your peer to win a promotion.

SUMMARY: RAISE THE MAGIC NUMBER

- *Reduce the return on politics*: Make lobbying for compensation and promotion decisions difficult. Find ways to make those decisions less dependent on an employee's manager and more independently assessed.
- *Use soft equity*: Identify and apply the nonfinancial rewards that make a big difference. For example: peer recognition, intrinsic motivators.
- *Increase project–skill fit*: Invest in the people and processes that will scan for a mismatch between employees' skills and their assigned projects. Adjust roles or transfer employees between groups when mismatches are found. The goal is employees stretched neither too much nor too little by their roles.
- *Fix the middle*: Identify and fix perverse incentives, the unintended consequences of well-intentioned rewards. Pay special attention to the dangerous middle-manager levels, the weakest point in the battle between loonshots and politics. Shift away from incentives that encourage battles for promotion and toward incentives centered on outcomes. Celebrate results, not rank.
- *Bring a gun to a knife fight*: Competitors in the battle for talent and loonshots may be using outmoded incentive systems. Bring in a specialist in the subtleties of the art—a chief incentives officer.
- *Fine-tune the spans*: Widen management spans in loonshot groups (but not in franchise groups) to encourage looser controls, more experiments, and peer-to-peer problem solving.

POSTSCRIPT

FROM NOBELS AND NUDGES TO NURTURING LOONSHOTS

A rapidly growing field called behavioral economics specializes in how incentives and environmental cues influence behavior. The influences studied by behavioral economists are often subtle, either because they are hidden or because they are based on quirks of psychology called cognitive biases. An example of a cognitive bias: In one study, experienced judges were asked to roll dice before sentencing. The jail terms they imposed were 60 percent longer after they rolled a high number than after they rolled a low number.

The judge example is disturbing, but it is in the context of a controlled experiment. A real-world disturbing example of a hidden influence at work is in the choice of delivery made by patients and physicians during childbirth. Since 1980, the rate of cesarean deliveries, called C-sections, has doubled in the United States. Today, nearly one in three births is delivered by C-section; it has become the most common surgical procedure in the country. The rate is far above the 10–15 percent range guided to by the World Health Organization and the Public Health Service (the surgery increases the risk to the mother of many serious complications). Recent studies by behavioral economists have shown that skewed financial incentives contribute to the excess. Physicians and hospitals are often paid more for C-sections than for vaginal births. The greater the difference, one study found, the higher the C-section rate. The results have led to a change in policies in some hospitals, requiring equal pay for both types of deliveries.

Requiring equal pay for both C-sections and vaginal births does not *tell* physicians and patients which treatment to choose, unlike, for example, a seat-belt law, which tells you to put on your seat belt. But it eliminates a perverse incentive. Simple changes that encourage, but don't mandate, behaviors we would like to see have been called "nudges." In their book with that title, Cass Sunstein and Richard Thaler offer a handful of policy examples, ranging from the serious (a plan that improves employee retirement-savings rates) to the less serious but equally effective (painting a fly on urinals has been shown to reduce urinal spillage by 80 percent). For his work in helping launch the field of behavioral economics, Thaler was awarded the 2017 Nobel Prize. For his work in bringing the psychology of individual decision-making—the study of cognitive biases—into economics, which inspired Thaler's work, Daniel Kahneman was awarded the 2002 Nobel Prize.

So what's the connection between these Nobels and nudges and the ideas of the previous chapters for nurturing loonshots more effectively?

What they have in common is a careful analysis of how incentives and environment can influence behavior, sometimes in hidden or unexpected ways. Where they differ is that, so far, behavioral economics has studied how environments influence *individual* decision-making. These past chapters describe how environments can influence *collective* decision-making—why teams and companies reject loonshots.

For example, judges ordering longer jail times after rolling a high number on dice *appears* to be irrational. But underneath that apparent irrationality is a rule about how the brain makes decisions that has evolved to help us complete ordinary tasks efficiently. (Sentencing criminals after rolling dice is not an ordinary task.) Similarly, a team rejecting valuable loonshots that everyone individually supports *appears* to be irrational. What we have done is explore the rational reasons why teams arrive at those decisions. In other words, why teams and companies, not just individuals, are "predictably irrational."

In both cases, understanding behavior can help us manage it. In one case, we may want to design environments that help individuals make better decisions. In the other, we want to help large groups innovate better.

The difference between studying the individual and studying the collective comes back to Phil Anderson's "more is different" motto. Anderson won his Nobel Prize for explaining why some materials can suddenly change from metals (good conductors of electricity) to insulators (bad conductors), called the metal-to-insulator phase transition. He also helped explain why some materials can suddenly change from ordinary metals to superconductors, in which all electrical resistance vanishes. Both are examples of *collective* behaviors. The electrons inside are the same. There is no way to understand either of these transitions by looking at the behavior just of *individual* electrons in isolation.

What we have been doing is combining Nobels: applying the principles of these two separate disciplines to the same problem. We are identifying how subtle changes in incentives can influence *collective* decision-making.

There is no way to understand why teams and companies suddenly change from innovating well to innovating poorly just by analyzing individual behaviors in isolation. The ability to innovate well is a *collective* behavior. It is another example of "more is different."

* * *

One final note. All of the above can be considered elements of *structure* in designing how individuals in teams or groups work together, in contrast to the previously mentioned mountains of print written about *culture.*

That a word or a subject has been abused to the point of meaninglessness, however, does not mean it should be dismissed entirely. Complex systems—the term of art for many interacting agents, whether buyers and sellers in markets, employees and managers in companies, or the atoms and molecules of a turbulent river—have earned that term for a reason. Their most interesting questions rarely have simple answers. In the complex system of our human body, for example, certain genes make diabetes or cancer more likely. But lifestyle choices also matter. Drinking sugary drinks by the gallon can bring on diabetes. Smoking cigarettes by the carton makes lung cancer more likely. *Both* genes and lifestyle matter. And so with teams and groups: *both* structure and culture matter.

The aim of this book is not to *replace* the idea that certain patterns of behavior are helpful (celebrating victories, for example) and others are less so (screaming), but to *complement* it.

In part one we saw how the lessons on structure from Bush and Vail, and an idea borrowed from a chess champion, can help us defeat chaos, stagnation, and the Moses Trap. In part two we saw how the science of phase transitions creates fresh insights into building more innovative groups. And we have seen how these ideas have come together, in a handful of examples, to win wars, cure diseases, and transform industries.

So now, to close, let's turn to another small topic: the history of our species.

PART THREE

THE MOTHER OF ALL LOONSHOTS

When you have learned to explain simpler things, so you have learned what an explanation really is, you can then go on to more subtle questions.

—Richard Feynman

9

Why the World Speaks English

THE NEEDHAM QUESTION

On a sunny day in August 1937, on the campus of Cambridge University in England, an attractive 33-year-old visitor knocked on the door of a renowned biochemist. Joseph Needham's three-volume study of how embryos form and grow had been compared by reviewers to Darwin's *Origin of Species*. Gwei-djen Lu had traveled two months and eight thousand miles, from Shanghai, to meet and possibly work with the legendary Dr. Needham and his wife, Dorothy Needham, an equally accomplished biochemist.

Lu expected "an old man with a bushy white beard." Instead, she found a man in his mid-thirties, tall and lean, whose strong voice had "a silkiness, almost a lisp" that she found mesmerizing. Lu soon learned that Needham had an uncommon range of interests: he was devoutly religious, an enthusiastic participant in group nude swimming, and an avid practitioner of free love. With Dorothy's knowledge, Needham began an affair with Lu.

Late one evening, several months later, according to his diary, Needham and Lu lay in bed, smoking cigarettes, when he suddenly turned to her. He asked if she would help him write the Chinese characters for "cigarette" in his diary. Together they wrote:

香煙

Needham studied the beautiful calligraphy, Lu recalled, then announced that he had to learn Chinese. She would be his teacher.

Needham's interest in the language soon extended to Chinese history. Western scientists and scholars, Lu had lectured him many times, failed to appreciate how many inventions and technologies had appeared first in China.

In the summer of 1942, Needham scribbled a note on a piece of paper: "Sci. in general in China—why not develop?" If so many ideas had appeared first in China, he asked Lu, why did the Scientific Revolution take place in Western Europe rather than China?

She had no answer. Needham decided to visit China, investigate the question, and summarize his findings in a short essay.

Needham never returned to biochemistry. Twenty-seven volumes, fifteen thousand pages, and three million words later—described by one reviewer as "perhaps the greatest single act of historical synthesis and intercultural communication ever attempted by one man"—Needham permanently changed the West's understanding of the East. He established exactly what Gwei-djen Lu had claimed: a vast number of technological, military, and political advances appeared first in China. In some cases, centuries earlier. In others, over a thousand years earlier.

But he never really answered the question he started with. Why the Scientific Revolution didn't take place in China, despite all its advantages, became known in world history circles as the Needham Question.

* * *

If you were an alien visitor from outer space, reading the history of the human species on earth like a novel, from its ape-like beginnings through the transformation from hunter-gatherers to domestic farmers, turning the pages, eagerly wondering when and where the revolution in science and industry might appear, you would almost certainly place your bets on China or India.

For a thousand years, from the middle of the first millennium AD to the middle of the second, China and India dominated the world's economy. Together, during this period, they averaged just over half of the world's GDP. The five largest nation-states of Western Europe, by comparison, averaged somewhere between 1 and 2 percent. Paper and printing appeared in China centuries before they appeared in Europe. The magnetic compass, gunpowder, cannons, crankshafts, deep-well drilling, cast iron, paper currency, sophisticated astronomical observatories: China. The imperial civil service exams—over a million tested annually, less than 1 percent passing—had been creating a class of scholar-elites in China for nearly a thousand years before the first universities opened their doors in Europe. Estimates of literacy rates in China around that time range as high as 45 percent. In England, it was close to 6 percent. In the early part of the fifteenth century, the Chinese navy sailed to North Africa and back with the largest fleet and ships ever seen—28,000 men and 300 ships, the largest weighing about 3,100 tons. A few decades later, Christopher Columbus sailed with three small ships, the largest weighing about 100 tons.

The Sino-Goliath was far larger, wealthier, and more technologically advanced than any of the far smaller Euro-Davids.

But something odd happened over the course of that long period. The Chinese giant turned inward, to big projects requiring massive resources. A new capital (Beijing). The Great Wall. The Grand Canal. Franchise projects. The Chinese leaders outgrew their interest in easily dismissed crazy ideas. The motion of planets, for example, or the properties of gases. Loonshots.

When the British approached China to expand trade in the eighteenth century, the Qianlong emperor wrote to King George III, "There is nothing we lack. We have never set much store on strange or ingenious objects, nor do we need any more of your country's manufactures."

Not long after, one of those strange and ingenious ideas arrived off the coast of China, powering the British ship *Nemesis*. Within weeks, the British fleet destroyed the old and outdated wooden sailing junks of the Chinese navy. The Chinese empire never recovered.

David's slingshot was the steam engine.

India during this period was ruled by the Mughal chiefs, heirs to a six-century reign of sultans and emperors. They also reveled in large franchise projects. The Taj Mahal, for example. Like the Chinese emperors, they passed on loonshots. In 1764, a private British trading company seized control of India. In 1857, India became a British colony.

Those strange and ingenious objects from Western Europe, which overpowered the much larger and wealthier empires of China and India, appeared as the result of a remarkable two-thousand-year cross-cultural journey—of Catholic bishops hiring Jews in Toledo to translate Arab critiques of Greek texts into Latin for Germans to read, of imported Chinese technologies and Indian mathematics and Islamic astronomy, of philosophers and popes, of eyeglasses and magnets and clocks and blood. That journey culminated in a new idea: underlying everything we see are universal truths that can be determined through measurement and experiment. In other words, laws of nature.

We take that idea for granted today. But for all human history up to that time, religious authorities or divine rulers or great-man philosophers decreed what was true and what was false. The idea that truth could be revealed to anyone was radical. Subversive. Its champions were often dismissed as unhinged.

That idea, now known by its more modern name, the scientific method, is arguably the mother of all loonshots.

The Chinese and Mughal emperors discovered the same lesson that surprised so many of their industrial descendants centuries later: missing loonshots can be fatal.

LOONSHOT NURSERIES IN INDUSTRY AND HISTORY

This book has been about creating conditions that encourage loonshots *inside* organizations. We can answer Needham's question—why Europe, and not China or India or anywhere else for that matter—by looking at how those principles apply *between* organizations. First, we'll see how loonshot nurseries form in industries, among companies. And then we'll extend that idea: we'll see how a loonshot nursery can form among *nations*.

We'll see why Western Europe, with its hundreds of independent city-states and small kingdoms, including England, was to the large empires of China and India what the teeming market of biotechs in Boston has been to Merck and Pfizer, and what the swarm of small production shops in Hollywood has been to Paramount and Universal. We'll see why Tycho Brahe succeeded, and why his equal and predecessor in China, five centuries earlier, came so close but didn't. We'll see why Western Europe became the flourishing loonshot nursery of its time—and what that means for nations today, who wish to avoid the fate of those ancient empires.

Let's start by taking a closer look at that mother of all loonshots.

EIGHT MINUTES THAT CHANGED THE WORLD

The path to the idea of laws of nature—and the scientific method for revealing those laws—mirrored, for good reason, the path to heliocentrism: the notion that the earth moves around the sun rather than the other way around. If divine rulers could be wrong about the most elementary questions of heaven and earth, then we needed a new way to define and seek truth.

The heliocentric idea first appeared in the fourth century BC, then periodically resurfaced, and was quashed, sometimes brutally, for nearly two thousand years. In the sixth century, the Indian astronomer Aryabhata suggested that the earth rotates about its axis every 24 hours, explaining the daily rotation of the stars and the sun in the sky. Hints of theories incorporating the motion of the earth appeared in both Christian Europe and outposts of the Islamic empire in the fourteenth and fifteenth centuries.

In Poland, in a small pamphlet completed around 1510 and privately circulated, Nicolaus Copernicus, a deeply religious Catholic church official, described in detail a system in which the earth revolves around the sun. He took pains to explain why his ideas posed no conflict with religion. The Vatican, intrigued, encouraged Copernicus to publish (conflict with the Church began only a century later, when Galileo ridiculed Church leaders). Copernicus resisted, sensitive not only to what his peers and other Church officials might think but also to his inability to answer the obvious flaws of his theory: If the earth spins about its axis every 24 hours at high velocity, why aren't birds flung from their nests? If we are hurtling around the sun, why isn't the moon left behind? In other words, like every loonshot, his theory arrived covered in warts.

Prodded by an eager disciple, Copernicus finally published three decades later, on his deathbed, in 1543. Few took his ideas seriously. Just as he had feared, the majority laughed at the warts and dismissed the whole thing. In 1589, the most prominent Italian astronomer, Giovanni Magini, wrote of Copernicus's ideas: "His hypotheses are rejected by practically everybody as being absurd." One historian identified only five scholars across all Europe around that time, five decades after Copernicus's death, who believed in his sun-centered world.

One of those five was a teacher at the University of Tübingen in

Germany named Michael Maestlin, whose lectures on planetary motion impressed a 17-year-old student named Johannes Kepler. This is Kepler describing himself in his diary:

> His appearance is that of a little lap-dog. His body is agile, wiry and well-proportioned. Even his appetites were alike: he liked gnawing bones and dry crusts of bread. . . .
>
> He is bored with conversation, but greets visitors just like a little dog; yet when the last thing is snatched away from him, he flares up and growls. . . . He hates many people exceedingly and they avoid him, but his masters are fond of him.

Kepler grew fascinated with Copernicus's ideas. He recognized the many unknowns and flaws. He understood that the theory was just as complicated as the ancient Greek system, with dozens of cycles and epicycles (circles upon circles) required to describe the orbits. And it was no more accurate—and therefore no more useful—than that widely used earth-centered system.

It was the sheer elegance of the idea that convinced Kepler, who was a bit of a romantic and more than a bit of a mystic. A sun-centered world explained much more naturally the inner planet motions (why Mercury and Venus never strayed far from the sun), as well as the unusual ordering of planetary periods. Planets closest to the sun complete their orbits quickly; those farthest from the sun take the longest.

The 24-year-old Kepler published a book filled with visions of giant pyramids and cubes in the sky shaping the orbits of the planets. He introduced his book with great enthusiasm: "For the first time I make this subject generally known to mankind . . . here we see how God, like a human architect, approached the founding of the world!" All his ideas were wrong. He later recanted many of them. But Kepler's brilliance as a mathematician shone through. He sent the book to Tycho Brahe, the leading astronomer in Europe, who immediately hired Kepler as an assistant. Tycho had his own theory of planetary motion, and he wanted young Kepler's help confirming it.

Tycho assigned Kepler the task of analyzing the motion of Mars. Kepler began his calculations by assuming circular motion, the only form considered perfect enough for heavenly objects. All prior planet-watchers—from the Babylonians to the Greeks, Arabs, and Europeans, up through Copernicus and including Tycho—began the same way. But despite five years of obsessive analysis, Kepler could not get rid of a small discrepancy between where his calculations predicted Mars should appear in the sky and what he saw looking through Tycho's instruments. It was an error of eight minutes of arc, less than one-twentieth of 1 percent. No matter how many and what form of cycles, epicycles, equants, and eccentrics he added (the mathematical tricks used by the Greek, Islamic, and European astronomers up to that time), he couldn't make that tiny difference go away. So Kepler decided to reject "that which exists only in the mind, and which Nature entirely refuses to accept": the assumption of circular motion.

Kepler's act lit a fire. In his *New Astronomy* (1609), Kepler announced, "Because they could not have been ignored, these eight minutes alone will have led the way to the reformation of all of astronomy."

In medicine or biology or zoology, the enormous variety of objects of study and the range of their behaviors leave little room for general laws. There is no universal theory of kidneys or of cats. Planets, however, repeat the same orbits, year after year, for millennia. A universal truth can be proposed—and carefully tested.

Kepler's ideas for universal truths were radical. The idea of elliptical orbits (even the idea of an orbit); the idea of a force from the sun that moves the planets; the idea that natural laws govern those motions; the idea that we should infer those laws from careful measurement—all of which Kepler introduced, all were new. Kepler broke far more violently from the past than Newton, who (mostly) unified then-existing principles with the goal of explaining Kepler's orbits. Kepler was closest in spirit to Einstein, three hundred years later, who also broke radically from the past. Einstein first rejected the ancient idea of an ether, a unique frame of reference in the universe to which everything should be compared. (His theory of special relativity said that the laws of physics are the same in any frame of reference; none is special.) Einstein then rejected Newton's action-at-a-distance gravity, the idea that a planet can exert a mysterious attractive force on an object far away. (His theory of general relativity explained those forces by showing how matter curves the space around it.)

Einstein saw in Kepler a "kindred spirit" who overcame religious persecution, poverty, personal tragedies, disbelieving audiences, and a heritage of mystical thinking. "Kepler's lifework was possible," Einstein wrote, "only

Kindred spirits: Albert Einstein and Johannes Kepler

once he succeeded in freeing himself to a great extent of the intellectual traditions into which he was born."

Unlike Kepler, Einstein benefited from a large and well-established scientific community. As mentioned earlier, the eclipse of 1919 confirmed Einstein's theory of gravity four years after he published it. Confirmation of Kepler's ideas proceeded much more gradually. In the decades after Kepler published his "War on Mars," astronomers and astrologers and navigators slowly realized Kepler's system worked far better than any earth-centric theory. Together with Galileo's discovery of the moons of Jupiter, William Gilbert's experiments with magnets, Robert Hooke's speculations on a universal gravity, and ultimately Newton's unifying laws, Kepler's radical ideas culminated in the widespread acceptance not only of a new astronomy but also a new way of thinking: truths judged by the outcome of experiments rather than the gavel of authority.

The rise and explosive spread of the scientific method across seventeenth-century Western Europe, in the decades after Kepler's death, and the revolution in the tools of industry it enabled sparked a pace and scale of change unlike any other in human history.

For ten thousand years, life expectancy barely changed. Between 1800 and 2000, it doubled. From AD 1 to 1800, global population grew less than 0.1 percent a year. By the mid-twentieth century it was growing at *20 times* that rate. The world's average economic output per person was nearly constant for two thousand years—between $450 and $650, in 1990 dollars. Since 1800, it has increased by a *thousand* percent.

The tiny nation-states of Western Europe, particularly England, rode that loonshot to global dominance—the principal reason the global language of business today is English rather than Chinese, Arabic, or Hindi.

THREE CONDITIONS FOR A LOONSHOT NURSERY

Which brings us back to Needham's question: why Western Europe?

Let's start by separating two questions that often get bundled together and shouldn't. The question of why some economies grew and others declined over the past two centuries often gets mixed with Needham's question. But Needham asks a *first-appearance*, creation question. Recent disparities are an *adoption* question: why did some countries adopt those new ideas of science and industry faster than others?

Haiti's economy, for example, declined over much of the twentieth century. The per-capita GDP of the Dominican Republic grew fivefold over the same period. Yet they are two halves of the same island. History doesn't allow proofs, but some explanations are not too difficult to *disprove*, or, at least, set aside. The natural experiment of Haiti and the Dominican Republic allows us to set aside differences in race, culture, climate, or geography—standard explanations for three centuries. Differences in political and economic institutions are much more natural explanations.

Needham's question is *not* about recent disparities like Haiti and the Dominican Republic. It is about a loonshot—the mother of all loonshots. Why did that loonshot appear and spread rapidly in Western Europe, in the seventeenth century, plus or minus a few years, when the empires of China, India, and Islam led the world in wealth, trade, organized study, and early science and technology for a *thousand* years?

The Islamic empire, for example, during various peaks from the ninth through the fifteenth century, exceeded both the West and China in the mathematics, astronomy, optics, and medicine, as well as in the libraries, hospitals, proto-universities, and observatories that gave birth to Western science. Copernicus borrowed many of his most critical mathematical steps directly from Arab astronomers. In 1025, the Persian physician and scholar Ibn Sina (called Avicenna in the West) wrote the *Canon of Medicine*. For seven centuries, it was the most widely used medical textbook in Europe.

We can again set aside the old explanations of culture, climate, and geography, just as we did for the *adoption* question. If Western European culture, climate, or geography were so much more favorable to progress than those in the lands of China, Islam, and India, then how could those ancient empires have dominated the world economy and technology innovation (paper, printing, magnetic compass, gunpowder, canal locks, advanced mining techniques, etc.) for so many centuries? Their culture didn't suddenly change. The height of their mountains didn't suddenly change.

For answers, as mentioned earlier, let's begin by looking at conditions that encourage loonshots within industries. The industries will show a familiar pattern: *phase separation* into two markets and *dynamic equilibrium* between those markets. We'll then apply that pattern to nations. We'll see why *structure* mattered more to the rise of the West, and the decline of the rest, than culture, climate, or geography.

* * *

For a loonshot nursery to flourish—inside either a company or an industry—three conditions must be met:

1. *Phase separation*: separate loonshot and franchise groups
2. *Dynamic equilibrium*: seamless exchange between the two groups
3. *Critical mass*: a loonshot group large enough to ignite

Applied to companies, the first two are the first Bush-Vail rules, discussed in part one. The third, critical mass, has to do with commitment. If there is no money to pay for hiring good people or funding early-stage ideas and projects, a loonshot group will wither, no matter how well designed. To thrive, a loonshot group needs a chain reaction. A research lab that produces a successful drug, a hit product, or award-winning designs will attract top talent. Inventors and creatives will want to bring new ideas and ride the wave of a winning team. The success will justify more funding. More projects and more funding increase the odds of more hits—the positive feedback loop of a chain reaction.

How many projects are needed to achieve critical mass? Suppose the odds are 1 in 10 that any one loonshot will succeed. Critical mass to ignite that reaction with high confidence requires investing in at least two dozen such loonshots (a diversified portfolio of ten of those loonshots has a 65 percent likelihood of producing at least one win; two dozen, a 92 percent likelihood).

To see how these three conditions apply to industries—*between* companies rather than *inside* a company—let's start with film. We'll see how the federal government helped separate the phases (#1).

MOVIES

Hustlers. In the early 1900s, young immigrants from Europe, scrap-metal collectors and fur traders and trinket peddlers with last names like Zukor, Mayer, Goldwyn, Loew, Cohn, Warner, Fox—mostly Jewish, some Catholic—jumped on Thomas Edison's new motion picture invention. They bought his equipment, rented his short films, and showed them in nickelodeons and penny arcades. In 1931, one writer compared those new moving pictures to the electric light, telephone, and even the steam engine:

> No other invention of the mechanical age had created such widespread astonishment and interest. . . . This new thing—this "living picture" affair was not a prosaic tool to reduce labor or to save time; it was not an instrument to create more comfort and luxury for the well-to-do. It was a romantic device to bring entertainment to the common people.

The hustlers rode that wave of wonder across the country, playing loose with Edison's patents along the way. They built theaters and hired writers, actors, and directors to make their own movies to fill those theaters. Edison tried to control or suppress them through his New Jersey–based patent company, hiring thugs to smash equipment and burn arcades. So they moved out West, near the Mexican border, where they could quickly flee with their pirate equipment whenever Edison's patent police showed up. They created a town of their own: Hollywood.

Over the next three decades, those penny-arcade cubs grew into studio-chief lions. Their studios—Paramount, Universal, MGM, Warner Brothers, Columbia—controlled everything from theaters to production lots to long-term contracts with talent. The oligopoly was glorious if you were a studio head, a Faustian bargain if you were a star, and a ripe target if you were a government antitrust lawyer. The Department of Justice began prosecuting the studios in the 1920s. It paused for the Depression, resumed, paused again for World War II, and finally, with the 1948 *US vs. Paramount* Supreme Court decision, broke apart Paramount and the other studios. No one producing movies could also own theaters.

The newly liberated market catalyzed an ownership circus. Studios were bought and sold by an automobile parts company, two beverage companies, a hotel company, a talent agency, a half dozen or so different conglomerates, and one Italian con artist working with a French government bank. The game of musical chairs peaked when Warner Brothers merged with Time Inc. and waltzed together into the arms of AOL, becoming the largest failed megamerger, at $186 billion, of all time.

As the music slowed, the film industry separated into two markets. The present-day Majors—Warner, Universal, Columbia, Fox, Paramount, and Disney—acquire and manage franchise or well-developed projects. They compete by delivering those products through as many channels to as many customers as they profitably can. They specialize in the scale and relationships needed to navigate opening night in New York, leveraged finance with Citibank, first release in Korea, on-demand with Netflix, video with Nintendo, toys with Walmart, theme-park land deals in Japan, and so on. On their quarterly earnings calls with analysts and investors—large mutual funds like Fidelity or T. Rowe Price—they discuss big-budget items like the future of the *Iron Man* franchise, or how measles outbreaks might affect theme-park revenues. Analysts speculate on next quarter's earnings and global market trends. The Majors don't discuss, and their analysts and investors aren't much interested in hearing about, recently acquired new scripts. Just as the Yankees don't discuss, and their fans and sportswriters aren't much interested in hearing about, double-A ballplayers in Trenton. The Major League markets specialize in acquiring and managing franchises.

The second market is a highly fragmented network of hundreds of small, independent production companies that bring together scripts, talent, and investors and herd them through the long dark tunnel to a finished film. They compete with their peers for access to new material, top creative talent, recognition at film festivals, and scarce funding. Fidelity and T. Rowe Price will never invest in them. Money comes from wealthy individuals and private money managers willing to gamble on crazy film projects, those that have been cast aside by the studio Majors. A metrosexual British spy, for example, who saves the world from evil men with long-range missiles and fluffy cats (James Bond). Or a boy from the slums of Bombay who appears on a quiz show (*Slumdog Millionaire*, eight Oscars). Or reptiles who love swords and pizza (*Teenage Mutant Ninja Turtles*, $1.2 billion in ticket sales). This is a market for creating, nurturing, and trading loonshots.

The industry survives and thrives because of the web of partnerships connecting the two markets (dynamic equilibrium, #2). Without the certainties of franchises, the high failure rates of loonshots would bankrupt the industry. But franchises grow stale. Without fresh loonshots, the large Majors would disappear.

Most of those partnerships are one-off deals. A small shop puts together a film, then solicits bids from the large Majors for rights to market it. Other partnerships are broader. Universal, for example, partnered with Imagine Entertainment for three decades and 50 pictures. Imagine found the stories and put together the films. Universal distributed. Their films together include the Oscar winners *Apollo 13* and *A Beautiful Mind*.

The two markets in film, connected by a web of partnerships, are examples of phase separation and dynamic equilibrium within an *industry*, rather than within a company. The market of hundreds of small production shops finding, funding, and developing small, crazy film projects is an example of an *industry's* loonshot nursery.

Government intervention—the breakup of the studio oligopoly—sparked phase separation in film.

In the biomedical world, the spark was a new technology.

DRUGS

A small number of large, global pharmas (Pfizer, Merck, Abbott, Roche, Eli Lilly) dominated drug development through the 1980s, in the same way that the old Hollywood studios had dominated the film industry through the 1940s. In both industries, product development begins by drawing on creative work done outside the industry. Stories in books or magazines provide starting material for movies (Ian Fleming's hero grew into the Bond franchise; *Flash Gordon* comics inspired *Star Wars*). Similarly, research from universities or national labs provides the starting points for new drugs. Cholesterol research from Konrad Bloch, Michael Brown, Joseph Goldstein, and others, for example, inspired the statins.

Up until the mid-1980s, that was the drug development industry: academics tended the wide field of research; the global pharmas drew on that research to create new drugs (production) and sell them to customers (distribution). Like the old Hollywood studio Majors, the big pharma Majors controlled both production and distribution. Until a young physician treating a 14-year-old boy created an entirely new kind of drug.

Most drugs we use come from nature—plants, animals, or microbes. The active ingredients in these natural-product drugs are relatively small molecules: aspirin, from willow tree bark, has just 21 atoms; morphine, from the opium poppy, has 40; Akira Endo's statin, from a mold, has 62. They fight disease by acting on proteins, the much larger molecules that do much of the work in a cell. When proteins malfunction, cells can spin out of control, causing disease. Natural-product drugs work by jamming into tiny crevices in overactive proteins, stopping them like a small wrench inserted into the guts of a giant, out-of-control robot. Aspirin blocks proteins involved in inflammation. Morphine blocks proteins that signal pain. Statins block a protein that regulates cholesterol levels. Chemotherapies block proteins (or other very large molecules) necessary for cell division. Nearly all drugs developed from the nineteenth through the late twentieth century are of this type.

The birth of a new kind of medicine began with a boy in Canada. On December 2, 1921, Leonard, 14 years old, was admitted to a Toronto hospital weighing 65 pounds, lethargic, his hair falling out, with acetone in his urine and dangerously elevated blood sugar. He was one of many children at the end stages of what is now called type 1 diabetes. State-of-the-art treatment was a starvation diet. Life expectancy was a few months. Twenty-five years earlier, a Polish-German scientist, Oskar Minkowski, discovered that removing the pancreas in animals caused symptoms of diabetes. Minkowski and many others tried administering ground-up animal pancreas as a treatment, but after more than 20 years of failed attempts, the leading American diabetes researcher wrote in a textbook, "All authorities are agreed . . . injections of pancreatic preparations have proved both useless and harmful. The failure began with Minkowski and has continued to the present without an interruption."

Meanwhile, a 29-year-old surgeon in Canada, with no experience in research and no funds (he supported himself by taking out tonsils and selling medical instruments), read an article about the pancreas. He grew curious and decided to work on the problem, either because he was courageous in the face of all those failures or—more likely—because he didn't read textbooks and was unaware of them. He came up with a new idea for extracting from the pancreas whatever mysterious substance might be controlling blood sugar. Working with a team in Toronto, he tried his preparation on some dogs and saw promising results. On January 11, 1922, he tried it on Leonard. The team waited anxiously in the hallway. Nothing happened. Their extract looked murky, so a biochemistry specialist was brought in to improve it. Twelve days later, Leonard was injected with a new mix.

Within 24 hours, Leonard's blood sugar fell almost 80 percent, and the acetone and sugar in his urine fell by almost 90 percent. He "became brighter, more active, looked better, and said he felt stronger," wrote the surgeon, Fred Banting, in a rapidly published medical report. The pancreas extract turned out to be a protein. Banting called it insulin. It saved Leonard's life.

Word of the new treatment spread quickly. The leading American diabetes researcher, Dr. Frederick Allen, flew to Toronto to secure a vial. One of his nurses wrote of the evening when he returned to the clinic with that vial:

> The mere illusion of new hope cajoled patient after patient into new life. Diabetics who had not been out of bed for weeks began to trail weakly about, clinging to walls and furniture. Big stomachs, skin-and-bone necks, skull-like faces, feeble movements, all ages, both sexes. It was a resurrection, a crawling stirring, as of some vague springtime. . . .
>
> I could see them drifting in, silent as the bloated ghosts they looked like. Even to look at one another would have painfully betrayed some of the intolerable hope that had brought them. So they just sat and waited, eyes on the ground.
>
> We all heard his step coming along the covered walk, past the entrance to the main hallways. His wife was with him, her quick tapping pace making a queer rhythm with his. The patients' silence concentrated on that sound.
>
> When he appeared through the open doorway, he caught the full

beseeching of a hundred pairs of eyes. It stopped him dead. Even now I am sure it was minutes before he spoke to them, his voice curiously mingling concern for his patients with an excitement that he tried his best not to betray.

"I think," he said,—"I think we have something for you."

Insulin changed medicine. Proteins were no longer just the *targets* of drugs; they could *be* drugs. Rather than block a misfiring robot with a tiny wrench, we replace the entire robot.

But there was a problem. Harvesting animal pancreases for every diabetic is no more practical than chopping down willow trees to make aspirin for every patient with fever. It took 50 years to find a solution. Developed in the 1970s, genetic engineering—which made it possible to grow mass quantities of purified human proteins in a lab—turned Banting's discovery of insulin into practical therapy.

Most of the big pharma Majors passed on the idea of a lab-grown protein as a new kind of medicine. The idea of engineered proteins as drugs was not too crazy, however, for a handful of entrepreneurs in the early 1980s, who started what became known as biotechnology companies. The success of their initial public offerings—most famously Genentech, described in chapter 5—established a market for a new type of company: one with no revenue, no profits, no sales force, and no certainty when, if ever, its technology would become a product. Those early entrepreneurs had created what was then, and is still today, a publicly traded market of loonshots.

If government intervention broke apart the Hollywood studio system, genetic engineering broke apart the pharma system. It separated production (the scientists who invent new drugs) from distribution (the pharmas that market them).

The pharma Majors are a small number of large multinationals (Novartis, Pfizer, Merck, Johnson & Johnson, Eli Lilly, etc.) with the scale and relationships to navigate product launches in Argentina, regulatory approval in France, manufacturing in Puerto Rico, leveraged finance with J. P. Morgan, reimbursement guidelines in Japan, and so on. On their quarterly earnings calls with analysts and investors—large mutual funds like Fidelity or T. Rowe Price—the Majors discuss big-budget items like the future of their cholesterol or diabetes franchises. Analysts speculate on next quarters' earnings and global market trends. The pharma Majors don't discuss, and their investors and analysts aren't much interested in hearing about, molecular pathways or early-stage drug candidates. It is a market for acquiring and managing franchises.

The investors and analysts who follow the hundreds of small companies in the biotech market, however, dive deeply into the science. The products are often still in laboratory studies or clinical trials, not yet approved by the FDA. There are no revenues to discuss—just biology, chemistry, and clinical trial data. The biotechs compete for starting material (technologies from universities or national labs), creative talent (biologists and chemists), and scarce funding from specialist investors. It is a market for unfashionable ideas routinely dismissed by the Majors. Gene therapies twenty years ago. Immunotherapies ten years ago. Stem cells today.

As in the film world, a symbiotic web of partnerships connects the two markets. Many are one-offs. In chapter 5, we saw how a one-off deal helped Genentech survive in its early days (a partnership with Eli Lilly), just like a one-off deal with Disney helped Pixar survive its early days. A handful of deals are much broader. The two-decade partnership between Roche, a pharma giant based in Switzerland, and Genentech, which is based in San Francisco, produced probably the greatest string of biotech hits seen in the industry so far. Those hits include the drug Avastin, mentioned in chapter 2, inspired by Judah Folkman's research, as well as the drug Herceptin that has transformed the treatment of breast cancer. Annual sales from their joint projects—the Swiss pharma's resources fueling the California company's loonshots—have exceeded $30 billion.

The hundreds of publicly traded or privately funded biotech companies are the loonshot nursery of the biomedical world.

* * *

Both the film and the drug-discovery industries have separated into two markets—the market of the Majors, who trade in franchises, and the market of small specialists, who nurture loonshots. Those two markets are connected by a web of partnerships. That separation and the connection are examples of the first two of the three conditions described earlier applied to industries: phase separation (#1) and dynamic equilibrium (#2).

The third condition, critical mass, is best illustrated with an example.

Nearly every major city in the United States over the past decade has come up with the idea of reinventing itself as a "biotech hub." Now suppose you are a recent biology or chemistry PhD. To which city are you more likely to relocate in search of a career: Detroit, which has a handful of biotech companies, or Boston, which is home to over two hundred companies along with hundreds of venture capitalists and biotech entrepreneurs launching dozens of new companies every year? Most biotechs, like most loonshots, struggle to survive. You want backup nearby in case your company goes belly-up. For similar reasons, the top investors, vendors, and suppliers flock to Boston. All want options and backup nearby. Recently, many of the world's largest biomedical companies have moved their research headquarters to Boston. They want to be close to companies and products they can acquire. More acquisitions means more venture dollars means more companies. It's a cycle that feeds on itself and grows—a virtuous cycle.

Boston has achieved critical mass and ignited. Detroit has not.

THE FATE OF EMPIRES

Now let's extend the three conditions to nations.

To begin, let's compare the fate of two men. One helped ignite the Scientific Revolution in Europe. The other, with similar ideas, similar approach, and similar—or greater—natural talent, might have done the same in China years earlier, but didn't.

Five centuries before Tycho Brahe built the best astronomical observatory in Europe, Shen Kuo assumed command of the best astronomical observatory in China. Tycho, a Danish nobleman, won support from the king of Denmark. The king awarded him the island of Hven and funds to hire a large staff and purchase the best equipment. Shen came from more modest birth and aced the imperial civil service exam. He studied astronomy in his spare time, rose through the ranks, and eventually won the support of the emperor, who appointed him head of the Imperial Bureau of Astronomy. (Astronomy was important to kings in Europe and emperors in China for similar reasons: signs from the heavens were interpreted as omens.)

Joseph Needham described Shen as "perhaps the most interesting character in all Chinese scientific history." Shen studied, wrote about, and in many cases contributed to a stunning breadth of fields: astronomy, mathematics, geology, meteorology, cartography, archaeology, medicine, economic theory, military strategy, anatomy, and ecology. He was the first to describe the magnetic compass and identify the difference between true north and magnetic north (which transformed navigation). He developed the earliest known examples of trigonometry and the mathematics of infinitesimals (the precursor to calculus) in China. He embodied what today we would call scientific curiosity. This is Shen wondering why lightning striking a house melts metal but leaves wood untouched:

> On certain wooden shelves, certain lacquered vessels with silver mouths had been struck by the lightning, so that the silver had melted and dropped to the ground, but the lacquer was not even scorched. Also a valuable sword made of strong steel had been melted to liquid, without the parts of the house nearby being affected.
>
> One would have thought that the thatch and wood would have been burnt up first, yet here were metals melted and no injury to thatch and wood.

Like Tycho, Shen wondered about the bizarre motions of the planets in the sky (mostly they drift eastward relative to the fixed background of stars, but for part of their orbits they appear to move backward, toward the west). Like Tycho, he insisted that only more accurate measurements could provide a deeper understanding. He designed the best astronomical measuring tools of his era, just like Tycho. Shen proposed to the emperor a program to measure the position of every planet, to high precision, three times a night every day for five years. Like Tycho, he hired brilliant assistants to complete his program (Tycho recruited Kepler; Shen recruited the blind mathematician Wei Pu).

Shen understood that his program was expensive. Funding required strong political support. He eventually lost that political support—just like Tycho. In Tycho's case, after King Frederick II of Denmark died, Tycho feuded with his son, the 19-year-old new king. Tycho wrote a letter to the young king, Christian IV, explaining exactly why he should continue to support Tycho's observatory and large staff. After all, Tycho was a famous European intellectual who brought glory to Denmark. The king replied that he was stunned by Tycho's "audacity and want of sense," and by the way that Tycho wrote "as if you were our equal." He cut Tycho's funding. Tycho lost his island and was forced into exile. In Shen's case, he was ousted from government, a casualty of similar political turnover and battles.

Shen Kuo: "I had only my writing brush and ink slab to talk to"

But here's the crucial difference: After Tycho left Denmark, he hunted around Europe for a new patron. King Rudolf II in Prague eventually raised his hand. Tycho moved his observatory there, which is where he brought Kepler and continued the work that ultimately led to Kepler's War on Mars and his "reformation of all of astronomy."

After Shen left government, on the other hand, he had nowhere to go. There were no other rulers who could support astronomy. And private support for astronomy was illegal—the study of the heavens was reserved for the emperor. So Shen spent the last decade of his life as a recluse, in exile, half of it under house arrest. His most famous work, spanning a dozen fields of study, was called *Brush Talks* because, Shen wrote, "since I retired and took residence in the woods, I have led a reclusive life and severed all social ties. Occasionally I recalled chats with my guests and put down one or two items with my brush . . . I had only my writing brush and ink slab to talk to."

When a script is killed inside Paramount or Universal or any studio Major, it stays dead. When an early-stage drug project is killed inside a major global pharma, it stays dead. In China—or in the various outposts of the Islamic empire—when the supreme ruler quashed promising new ideas about astronomy, as the emperor quashed Shen Kuo's ideas, they stayed dead.

Northern Song China of the eleventh and twelfth centuries, in Shen

Kuo's time, achieved critical mass. Steel and iron production grew explosively. Paper money, printing, and market exchanges proliferated. Song technology innovations spanned from the military (guns, cannons, bombs), to transportation (canals with pound locks), navigation (magnetic compass, sternpost rudders), and manufacturing (water-powered spinning for textiles). The period has been called "the first industrial miracle." The productivity and technological innovation were not matched until six centuries later in Europe.

Although China achieved critical mass (#3), it failed to ignite. It never created phase separation (#1) and dynamic equilibrium (#2). Political battles, and the emperor's own prejudices, would regularly override the conclusions of the early "scientists." Seven years after Shen began work on a new astronomical system, for example, the emperor decided it was good enough. He terminated the project and dismissed Shen's key assistant. It's as if Rudolf II had told Tycho his system was "good enough" and then fired Kepler.

The Song emperor failed to quarantine his loonshot group (phase separation) and maintain the balance between loonshots and franchises (equilibrium). In other words, he failed to do exactly what Vannevar Bush set out to do during World War II.

Another way to tell the same story, at the risk of historical and cultural whiplash: had the Song emperors appointed and listened to a Chinese Vannevar Bush, the scientific and industrial revolutions might well have taken place five centuries earlier. And we would all be speaking Chinese.

LOONSHOT LIFE SUPPORT

A critical role of the loonshot nursery is keeping fragile loonshots alive through failures and rejection.

In drug discovery and in the film industry, as mentioned earlier, projects killed inside the pharma or studio Majors generally stay dead (or become zombies: not quite dead, but not exactly living). In the loonshot nursery of small biotechs in Boston, however, or small production shops in Hollywood, a terminated project will float around just until a new investor raises his hand. For example, the most exciting new approach for treating cancer today—triggering the body's immune system to fight tumors—was rejected by every large pharma company. A handful of small biotechs, working closely with academics at universities and national labs, kept the idea alive. Most of those biotechs failed. A few succeeded, and they changed the treatment of cancer. The vast majority of the most important breakthroughs in drug discovery have hopped from one lily pad to another until they cleared their last challenge. Only after the last jump, from the final lily pad, would those ideas win wide acclaim.

When Tycho lost support from the king in Denmark and moved from one noble's castle to another for two years until he landed in Prague, he was similarly hopping from one lily pad to another. The flourishing loonshot nursery of local rulers willing to fund far-out research (and somewhat obnoxious researchers) not only rescued Tycho's observatory, but had also kept alive Copernicus's original idea of sun-centered orbits. A school in Wittenberg, Germany, taught aspects of Copernicus's system for six decades, despite its poor repute, until Tycho and Kepler finally rescued his theory.

But just the existence of a loonshot nursery—phase separation (#1)—is not enough. Eurocentric histories describing the rise of modern science in Western Europe often overlook the importance of the regular exchange with the large empires (dynamic equilibrium, #2). Without the mathematics borrowed from Indian scholars and Islamic astronomers, there would have been no Copernican theory. Without the navigation, transportation, communication, irrigation, mining, and military technologies imported from China, there would have been no surplus wealth or intellectual class in Europe to dream up theories of heavenly motions. All of which granted Western Europe the resources to achieve critical mass (#3).

And critical mass was an essential ingredient: overturning millennia of dogma required a string of loonshots, not just one. Some of those loonshots had appeared individually, much earlier, in other societies. The idea of planets orbiting the sun, as well as important precursors of calculus, appeared in the Kerala school in India centuries before Kepler and Newton. But as in China, those precursors failed to ignite. The critical mass in Europe, on the other hand, created a pan-European symphony of discoveries: telescopes (Netherlands), pointed at the sky (Italy), confirmed elliptical orbits (Germany) and the earth's motion (Poland), which were ultimately combined with ideas of inertia (Italy) and geometry (France) into a unified theory of motion (England). That's critical mass.

The empires of China, Islam, and India were the Majors of nation-states. The simmering stew of Western European nations was, at the time, the world's loonshot nursery for new ideas, just as the hundreds of small production shops serve as a loonshot nursery for new films, or the hundreds of small biotech companies serve as a loonshot nursery for new drugs.

The term *Majors* comes from sports. In baseball, the *Majors* refers to the league that features franchise players. Young talents are nurtured in the *Minors*. The terms vary, but most sports have a similar structure. What makes baseball unique is that the US Supreme Court has awarded baseball a special exemption from antitrust law. That exemption allows the Major League to control its membership, which keeps the Minor Leagues minor league.

In any industry *other* than baseball, Minors can grow up to be Majors. Disney began as a two-man shop (Walt and his brother), the tiniest of Minors. It built on the unexpected success of a mouse with big ears and a princess who befriends seven dwarfs to grow into one of the five studio Majors. Amgen began as a small biotech, as described earlier, a tiny Minor that came close to bankruptcy. It built on the astonishing success of its first drug to grow into a massive Major. Today Amgen has over $20 billion in annual sales.

Like every industry other than baseball, in the world of nation-states, a Minor can grow up into a Major. England began as a tiny Minor, just like Disney and Amgen. Just like those two, it built on the unexpected success of a powerful loonshot—the mother of all loonshots. It rode that idea to industrialize, weaponize, and evolve into a Major, spreading its language and customs around the world.

WHY ENGLAND?

We've been looking at the *global* first-appearance question: why did modern science appear first in Western Europe as opposed to the empires of China, Islam, or India? But there's another, more *local*, first-appearance question: why England as opposed to, say, France, Italy, or the Netherlands?

The answer cannot be a monopoly on brilliant scientists. Scientists in nearly every nation across Western Europe contributed crucial scientific steps, as described earlier.

Luck and timing always play a role in creativity and invention—the essence of a first-appearance story. Branch Rickey was the Hall of Fame baseball executive who created baseball's farm league system for developing new talent: players compete in the Minors and rise up to the Majors if they do well. He used that system to build eight World Series teams. It was Branch Rickey who originated the saying cited in part one: "Luck is the residue of design."

England did one thing quite differently—much better than its neighbors, which set it up to be luckier than its neighbors. England established the earliest example of a successful loonshot nursery *inside* one country.

The Royal Society of London, created in 1660, brought together nearly all the founders of modern science in England, including Robert Boyle, Robert Hooke, and Isaac Newton. It famously played a crucial role in helping and inspiring Newton. Without the Royal Society, as one historian noted, "It is doubtful that . . . there would ever have been a *Principia*." In other words, what we know today as Newton's laws most likely would go by some other name—or names. Gottfried Leibniz, for example, developed calculus independently, in Germany, around the same time as Newton. Christiaan Huygens, in the Netherlands, developed the idea of centripetal force, the wave theory of light, modern probability theory—and he invented the pendulum clock. Daniel Bernoulli in Switzerland, Leonhard Euler in Germany, Pierre-Simon Laplace in France—all were giants of mathematics and physics who arrived not long after Newton.

The Royal Society helped Newton and England win a race against time, a competition to discover truths of nature. But the Society didn't come together purely for basic research: "Science was to be fostered and nurtured as leading to the improvement of man's lot on earth by facilitating technologic invention."

In 1667, the Society's first historian and promoter, Thomas Sprat, wrote of "extraordinary Inventions" such as "Watches or Locks or Guns" and "Remed[ies] . . . against an Epidemical Disease" and declared that the "Publick should have Title to these Miraculous Productions." The purpose of the Royal Society, wrote Sprat,

> goes to the Root of all Noble Inventions and proposes an infallible Course to make England the Glory of the Western World.

Robert Hooke, Robert Boyle, and their air pump

A little more boldly stated than Vannevar Bush and his report. But it's the same basic idea, three centuries earlier.

As Sprat wrote those words, Robert Boyle was completing his experiments on the expansion and compression of gases, carried out by Hooke as his assistant. Hooke had built for Boyle what would soon become one of the most famous research devices in Europe: an air pump. Boyle used the device to discover the law now named after him (the pressure of a gas is proportional to its density).

After a few years working for Boyle, Hooke grew busy with his own work (inventing the microscope, proposing a universal gravity), so in 1675, Boyle hired a new assistant, a French medical doctor named Denis Papin. Papin continued the air-pump experiments, but added a twist. He was curious if he could add a piston to the pump and somehow create a working cycle of compression and decompression.

In 1687, Papin published a book describing how to use the Hooke-Boyle air pump to cook food. He called his new device a "Digester of Bones," since it squashed bones into edible bits. The 1687 book was a sequel to his first book, on the invention of what is now called a pressure cooker, so Papin called it *A Continuation of the New Digester of Bones.* Buried in the back, after a section on how to cook cows' horns and dried vipers, in what might be called the greatest example of burying the lead

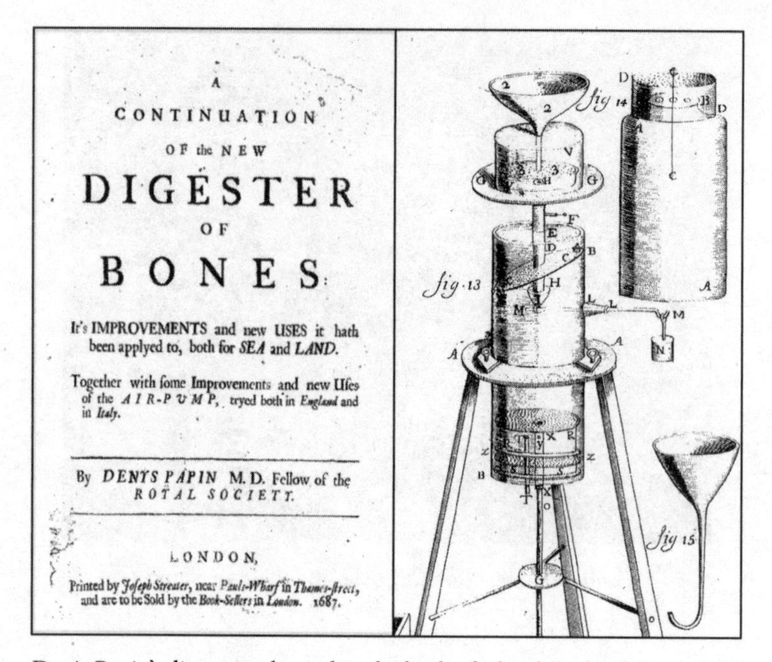

A
CONTINUATION
OF the NEW
DIGESTER
OF
BONES:
It's IMPROVEMENTS and new USES it hath been applyed to, both for *SEA* and *LAND*.

Together with some Improvements and new Uses of the *AIR-PUMP*, tryed both in *England* and in *Italy*.

By *DENYS PAPIN* M.D. Fellow of the *ROYAL SOCIETY*.

LONDON,

Printed by *Joseph Streater*, near *Pauls-Wharf* in *Thames-street*, and are to be Sold by the *Book-Sellers* in *London*. 1687.

Denis Papin's discovery, buried in the back of a book on kitchen utensils

in history, was the answer to his puzzle on how to add a piston to Boyle's air pump. It described the key components for a new invention: a steam engine.

Although the scholars of the Royal Society paid little attention to Papin's ideas, especially since they appeared in the back of a book about cooking, those ideas did not escape the notice of a craftsman in Dartmouth, England, named Thomas Newcomen. Newcomen had little interest in philosophy but a lot of time for useful gadgets, like pressure cookers.

In 1712, Newcomen turned Papin's movable piston inside a pump into the first practical, workable steam engine. Newcomen's invention rapidly spread throughout England. Over the next century, inventors continued to improve its efficiency. The engine soon elevated production of resources and goods far, far past the limits set by human or animal power, the limits that had held human societies around the globe to a fixed level of production for thousands of years. The change, which began in England and soon spread to the rest of Europe, fueled Western Europe's rapid rise to global power, the defeat of much larger and older empires, and an exponential growth in human population.

* * *

The Royal Society of London; Vannevar Bush's wartime loonshot nursery, the OSRD; and Theodore Vail's Bell Labs—all three had something in common. They were the greatest loonshot nurseries of their time. They were, arguably, the three greatest loonshot nurseries in history. They produced the Scientific Revolution, victory in a world war, and the transistor.

Why did the empires of China, Islam, and India miss the Scientific Revolution despite their wealth and historical advantages? For the same reason that Microsoft missed mobile, Merck missed protein drugs, and the film Majors missed *My Big Fat Greek Wedding.* Loonshots flourish in loonshot nurseries, not in empires devoted to franchises. Being good at loonshots and good at franchises are phases of an organization—whether that organization is a team, a company, or a nation. That's what the science of emergence tells us.

To survive the next revolution—whatever that might be—nations and their leaders should heed the lessons of Vannevar Bush and Theodore Vail. Some can be found in earlier chapters, in the lessons for teams and companies applied to nations. Many may be found in the *Endless Frontier* report of Vannevar Bush, written in 1945, at the request of President Roosevelt.

FDR wrote to Bush, "New frontiers of the mind are before us, and if they are pioneered with the same vision, boldness, and drive with which we have waged this war we can create a fuller and more fruitful employment and a fuller and more fruitful life."

With a little help, and a little science, we can each press on, as individuals, as members of teams, as citizens of nations, toward our own endless frontiers.

Afterword

Loonshots vs. Disruption

This afterword is mostly for business-theory or innovation-theory junkies who may have heard of, or even occasionally make use of, the term *disruptive* or the term (which causes me even more stomach pain) *disruptive innovation.*

First, to get something out of the way quickly: the two types of loonshots described in chapter 3 are unrelated to what Louis Galambos in 1992 called "adaptive" vs. "formative" innovations, and Clayton Christensen in 1997 called "sustaining" vs. "disruptive" innovations. The two loonshots distinguish between a new strategy (S-type) and a new product or technology (P-type). Galambos and Christensen distinguish between improvements to existing products (sustaining) and technologies that eventually significantly alter some market (disruptive). Christensen specifically emphasizes new products, from new entrants, that begin in the "low end" of a market, with inferior quality, and gradually improve until they win over the high-end customers of an incumbent.

S-type and P-type loonshots could each be either disruptive or sustaining, by these definitions. And vice versa: disruptive and sustaining innovations could each be either S-type or P-type. These describe different, unrelated properties, like height and hair color.

LOONSHOTS (TODAY) VS. DISRUPTIVE (IN HINDSIGHT)

A loonshot refers to an idea or project that most scientific or business leaders think won't work, or if it does, it won't matter (it won't make money). It challenges conventional wisdom. Whether a change is "disruptive" or not, on the other hand, refers to the effects of an invention on a market.

This book is about the former, not the latter, because, as experienced entrepreneurs know, so many ideas and technologies now recognized as transformative began with practically no resemblance to the final product they grew into, nurtured by champions who never imagined their ultimate market. Early-stage projects in rapidly evolving markets behave like a leaf in a tornado. You wouldn't put a lot of faith in guessing where that leaf might end up.

It's easy to point to technologies that disrupted a market in *hindsight*, once the leaf has landed. We know that the transistor launched the electronics age. We know that personal computers can empower individuals and replace mainframes or minicomputers. We know that Walmart grew astronomically, and competitors disappeared. We know that biotechnology produces important drugs. But what about when those ideas first took shape?

THE TRANSISTOR

Could scientists at Bell Labs in the 1940s working on the band theory of solids, or germanium semiconductors, or the science of surface states have said that they were working on a disruptive technology? They were given the vague goal of improving the performance of existing amplifiers and relay switches used in the phone system. By the definitions above, their goals were *sustaining.*

Even several years after the invention of the point-contact transistor in 1947, no one quite knew what to do with it. The first commercial application, inside hearing aids, did not appear until late 1952. Did the scientists or business managers working on the transistor begin with the idea of disrupting the hearing-aid market? No. They were building better switches.

Did the transistor come from a new entrant, start off low-priced, for the low end of a market? No. It began as a sustaining innovation from the *largest* company in the country. It was initially much *more* expensive than a vacuum tube ($20 vs. $1). It first sold to *high-end* customers like the military.

Later, of course, the transistor got cheaper and disrupted nearly every market.

ONLINE SEARCH

To fast-forward a few decades: could Google, when it began, say that it had developed a disruptive innovation? Larry Page and Sergey Brin's improved algorithm for prioritizing internet search results, PageRank, was incrementally more helpful to users than results from the many other existing search engines. It was a "sustaining" innovation, by the definitions above.

WALMART

When Sam Walton opened stores in rural areas, far from big cities, was he thinking it might be a strategic, disruptive innovation?

"Man, I was all set to become a big-city department store owner," he wrote about opening his first store. He was looking at St. Louis. "That's when Helen spoke up and laid down the law." His wife announced, "I'll go with you any place you want so long as you don't ask me to live in a big city. Ten thousand people is enough for me." He ended up in Bentonville, Arkansas, population: 3,000, in part because "I wanted to get closer to good quail hunting, and with Oklahoma, Kansas, Arkansas, and Missouri all coming together right there it gave me easy access to four quail seasons in four states."

The result was the leaf in the tornado.

"It turned out that the first big lesson we learned," wrote Walton, years later, "was that there was much, much more business out there in small-town America than anybody, including me, had ever dreamed of."

IKEA

While we're on retail, let's talk about furniture.

In 1948, a 22-year-old Swede named Ingvar Kamprad, with a small mail-order business selling Christmas cards, pens, picture frames, and the like, added furniture to his list. He advertised items from local designers. His business grew enough to threaten larger Swedish furniture-store owners. They had him banned from exhibiting at the usual trade fairs (a carpet-dealer friend once smuggled him into a fair in the back of a Volvo by throwing a carpet over him).

In response, Kamprad filled a large, empty warehouse in the Swedish countryside with samples of his furniture for customers to see before they ordered off his list. That was the first IKEA showroom. An employee trying to stuff a table into his Volvo realized he could save space by removing the legs and storing them under the table. Because shipping costs were rising, they decided to try the same trick in shipping to customers. Customers went for it, and self-assembly was born. Orders grew.

Furniture-store owners retaliated by forbidding designers to work with Kamprad. He was forced to hire his own designers. That led to original IKEA brands and style—furniture you own but can't pronounce: Poäng, Alvangen, Grundvattnet.

Once Kamprad began building his own furniture, the store owners banned their wood suppliers and other manufacturers from working with him. So Kamprad went to Poland and discovered high-quality suppliers—for half the price. He passed the discounts on to customers. Business, of course, grew. Years later Kamprad wrote, "Who knows whether we would have been as successful as we were if they [the Swedish furniture manufacturers] had offered us an honest fight?"

In 1965, IKEA opened its first store in Stockholm. There were so many customers that the store manager let customers go directly to the warehouse at the back of the store and take their own items. Which gave birth to self-service warehouses. All future stores were designed to allow customers to shop the warehouse.

In 2017, IKEA annual sales exceeded $44 billion. Visits to the 403 stores in 49 countries reached nearly one billion.

None of the defining elements of what became the planet's largest furniture store began with the idea of "disrupting" an industry. They were all small, crazy ideas explored by Kamprad and his team, in a desperate bid to survive.

THE TRUTH ABOUT DRUG DISCOVERY

In drug discovery, market estimates for early-stage products are notoriously unreliable, almost comically so, in hindsight. Amgen's drug for increasing the body's production of red blood cells, mentioned in the introduction, was expected to help only a small fraction of kidney disease patients, those whose kidneys produce too few red blood cells. Nearly every major pharma company considered, and rejected, offers to acquire the technology or the company, because the market projections were so small. At one point, Amgen nearly ran out of cash. Later, researchers discovered that the same drug helps cancer patients treated with chemotherapy, who also suffer from low blood cell counts. Millions of patients benefited from the drug. Amgen grew into a $100 billion company.

In the early 1980s, scientists and the public became fascinated with another drug that might help cancer patients, called interferon. In laboratory experiments, the drug seemed to interfere with the activity of viruses, giving rise to its name and the hope that it might be a magic bullet against infectious disease. Trials proved disappointing, however. The idea that tumors were caused by viruses was popular at the time, so a handful of researchers decided to test whether interferon could help treat cancer. Headlines trumpeted early results: "Magic Drug Saves Boy! Interferon Works Wonders!" Later trials in cancer, as with infectious disease, proved disappointing. Interest quickly faded.

Years later, a handful of scientists discovered that the drug does work surprisingly well: for treating multiple sclerosis. To this day, researchers don't know why interferon works in MS. Scientists began the trials because they thought MS might be caused by a virus. As with cancer, that turned out not to be the case. Still, the drug works. The product grew into a franchise with over $6 billion in annual sales. Could scientists working on interferon have declared they were working on a "disruptive technology"? For what market?

The multiple sclerosis need turned out to be much bigger than anyone imagined, but not nearly as big as the need in rheumatoid arthritis (RA). Nearly all the major pharma companies dismissed new drugs for treating RA when they were first developed, because RA was considered an "old lady" disease, a tiny market. Today the leading category of drugs for treating RA sells just over $30 billion annually. It turned out that severe

RA was just one of a broad range of serious autoimmune disorders, including Crohn's disease, psoriasis, ulcerative colitis, and a handful of others that could be treated effectively with the new drugs.

A few years ago, I had lunch with the recently appointed CEO of a major pharma company. As he rose up through the ranks, he never lost a skeptical view of marketers. When the topic of estimating the market for some new drug came up, he told me a story. When he was first appointed CEO, he asked his marketing group to prepare a summary of their 20 most recent product launches and calculate for how many the actual sales fell within a factor of two of the original sales projection. The answer: zero.

So what do these stories, and the stories from earlier in the book, tell us?

USE "DISRUPTIVE INNOVATION" TO ANALYZE HISTORY; NURTURE LOONSHOTS TO TEST BELIEFS

In an article addressing recent controversy about the notion of disruptive innovation, Christensen explains why Uber is not disruptive, by his definition, and why the iPhone also began as a sustaining innovation. In chapter 3, we saw that American Airlines—a large incumbent, not a new entrant—led the airline industry after deregulation with many brilliant "sustaining" innovations targeted to high-end customers. Hundreds of low-cost, specialty airline startups, "disruptive innovators," failed.

If the transistor, Google, the iPhone, Uber, Walmart, IKEA, and American Airlines' Big Data and other industry-transforming ideas were all initially sustaining innovations, and hundreds of "disruptive innovators" fail, perhaps the distinction between sustaining vs. disruptive, while interesting academically or in hindsight, is less critical for steering businesses in real time than other notions.

That, at least, is why I don't use the distinction in this book. I use the distinction between S-type and P-type because teams and companies or any large organization develop deeply held beliefs, sometimes consciously, often not, about both strategies and products—and loonshots are contrarian bets that challenge those beliefs. Perhaps everything that you are sure is true about your products or your business model is right, and the people telling you about some crazy idea that challenges your beliefs are wrong. But what if they aren't? Wouldn't you rather discover that in your own lab or pilot study, rather than read about it in a press release from one of your competitors? How much risk are you willing to take by dismissing their idea?

We want to design our teams, companies, and nations to nurture loonshots—in a way that maintains the delicate balance with our franchises—so that we avoid ending up like the Qianlong emperor. The one who dismissed those "strange or ingenious objects," the same strange and ingenious objects that returned in the hands of his adversaries, years later, and doomed his empire.

ACKNOWLEDGMENTS

I am enormously grateful to the many friends and experts who have helped shape this book through patiently answering dozens of questions, generously reading and rereading, and especially for delicately explaining to me, on occasion, why something I thought was funny was not quite so much.

For decades of world-class scientific mentorship and friendship: Lan Bo Chen, Steve Kivelson, Bob Laughlin, Bryan Lynn, Lenny Susskind, as well as the late Sir James Black and Judah Folkman. For advice and friendship in business and finance: there are so many fine people who put in so many hours and so much heart into so many projects together, and from whom I have learned so much, that I cannot possibly do the list justice. You know who you are, and I thank you for everything.

For their valuable comments, practical writing advice, and help with specific subjects in *Loonshots*: Christopher Bonanos, Christa Bourg, Dorie Clark, Susan David, Masaki Doi, Iman El-Hariry, Akira Endo, Alex Farman-Farmaian, Ben Feder, Josh Foer, Owen Gingerich, Matt Gohd, Joseph Goldstein, Nir Hacohen, Sue Halpern, Ken Howery, Neil Johnson, Manolis Kellis, Jonathan Leaf, Jennifer 8. Lee, Nimitt Mankad, Art McMahon, Doug Miller, Richard Miller, Robert Montgomery, Scott O'Neil, Richard Preston, Beth Rashbaum, Susan Schmidt, Kim Scott, Nathan Sivin, David Spergel, Steven Strogatz, Becky Sweren, Lauren Terry, Philipp Thornquidt, Ed Trippe, Greg Warner, Alex Wellerstein, Doug

Wickert, and Akira Yamamoto. Three heroes made it all the way through endless drafts and offered priceless suggestions at every stage: Paul Craig, John Thompson, and especially Andrew Wright. I am grateful for the many wise advisors. Inclusion on this list signals only my gratitude, not agreement or endorsement. All errors and omissions are my own.

Special thanks to Bill Press for recruiting me for PCAST; Zic Rubin for legal advice; Paul Craig for many years as my gracious New York host and thoughtful sounding board; Olivia Fox Cabane and Dan Ariely for recommending Jim Levine; Tim Harford for recommending Andrew Wright; and Dina Kraft for recommending Chia Evers. Had I found Chia earlier, this would have been a better book. She applied her superb research skills to chasing down the obscure fact; finding the obscure reference; and organizing and pruning over 5,000 heavily annotated journal articles, book chapters, and news sources across dozens of unconnected disciplines into a sensible bibliography, all of which included deciphering years of my notes. I have no idea how she did all of that without ever dropping her smile or losing her cool. I am forever grateful.

Freeman Dyson encouraged me from the beginning ("Contrary to the predictions of the experts, books are still an effective way to spread ideas"). In his *Origins of Life*, Dyson gently introduces an original toy model as a new way of thinking about a long-standing problem. As he has done for his entire career, he does so without regard to what experts might think of his intrusions into their fields. Which inspired me to do the same.

The magnificent scratchboard engraving illustrations throughout the book are the work of Antar Dayal. Antar transformed vague ideas and stick-figure drawings into wildly inventive, beautifully stylized visions, bursting with life and humor. Antar's style of collaborating matches his style of art, both of which livened up every day.

At St. Martin's Press, Tim Bartlett has lived up to his reputation as an outstanding editor, pruner, untangler, and gentle suggester—a master of the micro and the macro. Every writer I spoke to told me I was lucky to work with him. They were right. Thanks also to the rest of the team at SMP, especially Alan Bradshaw, Laura Clark, Katherine Haigler, and Alice Pfeifer, white-gloved experts in caring for both manuscripts and authors.

Jim Levine patiently coaxed a complete book out of a tiny essay, wisely guided me through unfamiliar territory, and reminded me, exactly when I needed to be reminded, of the joy and wonder of big ideas and great stories. Thanks also to Elizabeth Fisher and Matthew Huff, who work with Jim at the LGR agency, for their gracious help.

Bahcall family members have provided the support and encouragement that made all of this possible. To Dan, Orli, and my mother, I owe so much. My mother, Neta Bahcall, provided valuable input on the science, especially the history of astronomy, read dozens of drafts, and only occasionally asked me about my prospects for a real job.

Words cannot describe how much Ethan, Julia, and Magda have transformed and enriched my life. The book is dedicated to my father, my inspiration. I am dedicated to Magda, my guiding light, and our life and children together.

GLOSSARY

Terms invented for this book are indicated with an asterisk (*)

Complex system
A whole made of many interacting parts whose interactions follow certain rules or principles.
Examples: (1) **Water** is made of many molecules that attract and repel each other through electromagnetic forces. (2) **Traffic flow** is made of many drivers who seek to achieve a target cruising speed while not crashing into other drivers. (3) **Markets** are made of buyers who want the best products at the lowest prices, and sellers who want to make the greatest profit.

Control parameter
A variable that can alter the state of a complex system
Examples: (1) **Water**: A small change in temperature or pressure can trigger a transition from a solid to a liquid phase. (2) **Traffic flow**: A small change in the density of cars on a highway or the average car speed can trigger a switch from smooth flow to jammed flow. (3) **Markets**: The degree of influence that one buyer's behavior has on another, *herding*, can change a market. High herding encourages a bubble phase; low herding discourages a bubble phase.

Dynamic equilibrium
When two phases coexist in balance, continuously exchanging their parts, neither side growing or shrinking at the expense of the other. For example, when blocks of ice coexist with pools of water as molecules cycle back and forth between the two.

Emergent behavior (or property) A property of the whole that cannot be defined or explained by studying the parts on their own. The behavior emerges from how those parts interact *collectively* rather than what they do *individually.*

Examples: (1) **Water:** Ice is rigid and shatters when struck; liquid water is slippery and flows when poured. That behavior cannot be defined or explained by studying a water molecule on its own. (2) **Traffic flow:** Cars on a highway may flow smoothly with no interruption or they may jam in response to small disruptions. Those emergent behaviors don't depend on the details of the cars or the drivers. (3) **Markets:** Prices adjust to demand and resources tend to be allocated efficiently, regardless of what buyers are buying or sellers are selling.

Unlike fundamental laws, emergent behaviors can suddenly change. When monopolies or cartels appear in a market, for example, prices may no longer adjust to demand and resources may no longer be allocated efficiently.

False Fail* When a valid hypothesis yields a negative result in an experiment because of a flaw in the design of the experiment.

Franchise The subsequent iterations or updated versions of an original product or service. Examples: the ninth statin drug; the twenty-sixth James Bond movie; the iPhone X.

Life on the edge* Life on the edge of a phase transition: when a control parameter brings a complex system to the cusp of a transition. Example: adjusting the temperature of water to 32 degrees Fahrenheit. Phases will separate and coexist in dynamic equilibrium.

Loonshot* A neglected project, widely dismissed, its champion written off as unhinged.

Moonshot An ambitious and expensive goal, widely expected to have great significance. A moonshot is a destination (for example, the goal of eliminating poverty). Nurturing loonshots is how we get there.

Moses Trap* When an all-powerful leader becomes judge and jury deciding the fate of loonshots.

Phase A state of a complex system characterized by a specific set of emergent behaviors.

Examples: (1) **Water** molecules can arrange themselves into a rigid, ordered lattice (solid phase) or they may bounce around randomly (liquid phase). (2) **Traffic flow:** small disruptions may grow exponentially into a jam (jammed flow) or they may have no effect (smooth flow). (3) **Markets:** buyers may respond mostly to some estimated fair value of a seller's product (rational phase) or they may respond mostly to what other buyers are doing (bubble phase).

Phases of organization* When an organization is considered as a complex system, we can expect that system to exhibit phases and phase transitions—for instance, between a phase that encourages a focus on loonshots and a phase that encourages a focus on careers.

Phase transition A sudden change between two phases, i.e., between two types of emergent behaviors.

Examples: between solids and liquids; between smooth flow and jammed flow on highways; between rational and bubble phases in markets.

P-Type loonshot* A new *product* or technology that no one thinks will work.

S-Type loonshot* A new *strategy* or business model that no one thinks will achieve its goal.

APPENDIX A
SUMMARY: THE BUSH-VAIL RULES

1. **Separate the phases**
   - Separate your artists and soldiers
   - Tailor the tools to the phase
   - Watch your blind side: nurture *both* types of loonshots
2. **Create dynamic equilibrium**
   - Love your artists and soldiers equally
   - Manage the transfer, not the technology: be a gardener, not a Moses
   - Appoint and train project champions to bridge the divide
3. **Spread a system mindset**
   - Keep asking *why* the organization made the choices that it did
   - Keep asking *how* the decision-making process can be improved
   - Identify teams with outcome mindset and help them adopt system mindset
4. **Raise the magic number**
   - Reduce return-on-politics
   - Use soft equity (nonfinancial rewards)
   - Increase project–skill fit (scan for mismatches)
   - Fix the middle (reduce perverse incentives for middle managers)
   - Bring a gun to a knife fight (engage a chief incentives officer)
   - Fine-tune the spans (wide for loonshots groups; narrow for franchise groups)

For anyone championing a loonshot, anywhere:

- Mind the False Fail
- Listen to the Suck with Curiosity (LSC)
- Apply system rather than outcome mindset
- Keep your eyes on SRT: spirit, relationships, time

The first three rules are discussed in part one, chapters 1 through 5. The fourth rule is discussed in part two, chapters 7 and 8.

1. Separate the phases

- *Separate your artists and soldiers:* Create separate groups for inventors and operators: those who may invent the next transistor vs. those who answer the phone; those who design radically new weapons vs. those who assemble planes. You can't ask the same group to do both, just like you can't ask water to be liquid and solid at the same time.
- *Tailor the tools to the phase:* Wide management spans, loose controls, and flexible (creative) metrics work best for loonshot groups. Narrow management spans, tight controls, and rigid (quantitative) metrics work best for franchise groups.
- *Watch your blind side:* Make sure your loonshot nursery seeds both types of loonshots, especially the type you are least comfortable with. S-type loonshots are the small changes in strategy no one thinks will amount to much. P-type loonshots are technologies no one thinks will work.

2. Create dynamic equilibrium

- *Love your artists and soldiers equally:* Artists tend to favor artists; soldiers tend to favor soldiers. Teams and companies need both to survive and thrive. Both need to feel equally valued and appreciated. (Try to avoid calling one side "bozos.")
- *Manage the transfer, not the technology:* Innovative leaders with some successes tend to appoint themselves loonshot judge and jury (the Moses Trap). Instead, create a natural process for projects to transfer from the loonshot nursery to the field, and for valuable feedback and market intelligence to cycle back from the field to the nursery. Help manage the timing of the transfer: not too early (fragile loonshots will be permanently crushed), not too late (making adjustments will be difficult). Intervene only as needed, with a gentle hand. In other words, be a gardener, not a Moses.

- *Appoint and train project champions to bridge the divide:* Soldiers will resist change and see only the warts on the baby-stage ideas from artists. Artists will expect everyone to appreciate the beautiful baby underneath. They may not have the skills to convince soldiers to experiment and provide the feedback that is crucial for ultimate success. Identify and train bilingual specialists, fluent in both artist-speak and soldier-speak, to bridge the divide.

3. Spread a system mindset

- *Keep asking* why: Level 0 teams don't analyze failures. Level 1 teams assess how product features may have failed to meet market needs (outcome mindset). Level 2 teams probe *why* the organization made the choices that it did (system mindset). They analyze *both* successes and failures because they recognize that good outcomes don't always imply good decisions (got lucky), just as bad outcomes don't always imply bad decisions (played the odds well). In other words, they analyze the quality of *decisions,* not just the quality of *outcomes.*

- *Keep asking* how *decision-making processes can be improved:* Identify key influences—people involved, data considered, analyses conducted, how choices were framed, how market or company conditions affected that framing—as well as both financial and nonfinancial incentives for individuals and for the team as a whole. Ask how those influences can be changed to enhance the decision-making process in the future.

- *Identify teams with outcome mindset and help them adopt system mindset:* Analyzing a product or a market may be technically challenging, but it is a familiar and straightforward exercise. Analyzing *why* a team arrived at a decision can be both unfamiliar and uncomfortable. It requires self-awareness from team members; the self-confidence to acknowledge mistakes, especially interpersonal ones; and the candor and trust to give and receive delicate feedback. The process is likely to be more efficient, and less painful, when it is mediated by a neutral expert from outside the team.

4. Raise the magic number

- *Reduce return-on-politics:* Make lobbying for compensation and promotion decisions difficult. Find ways to make those decisions less dependent on an employee's manager and more independently assessed and fairly calibrated across the company.
- *Use soft equity:* Identify and apply the nonfinancial rewards that make a big difference. For example: peer recognition, intrinsic motivators.
- *Increase project–skill fit:* Invest in the people and processes that will scan for a mismatch between employees' skills and their assigned projects, and will help managers adjust roles or employees transfer between groups. The goal is to have employees stretched neither too much nor too little by their roles.
- *Fix the middle:* Identify and fix perverse incentives, the unintended consequences of well-intentioned rewards. Pay special attention to the dangerous middle-manager levels, the weakest point in the battle between loonshots and politics. Shift away from incentives that encourage battles for promotion and toward incentives centered on outcomes. Celebrate results not rank.
- *Bring a gun to a knife fight:* Competitors in the battle for talent and loonshots may be using outmoded incentive systems. Bring in a specialist in the subtleties of the art—a chief incentives officer.

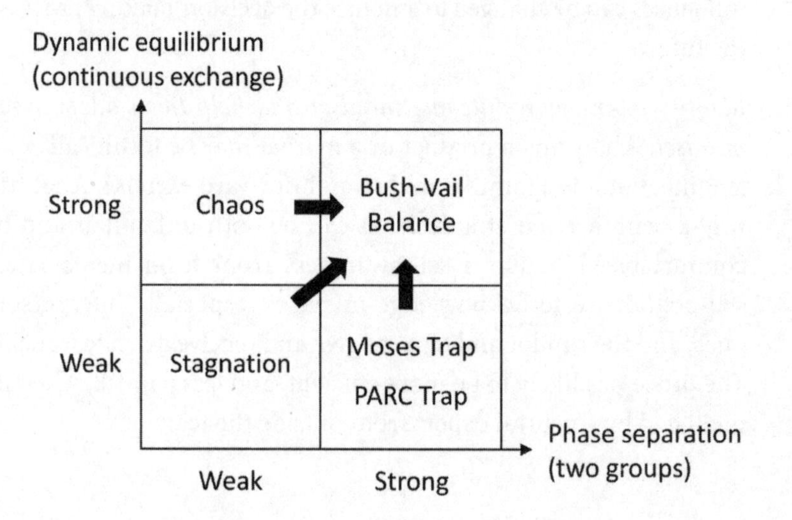

- *Fine-tune the spans:* Widen management spans in loonshot groups (but not in franchise groups) to encourage looser controls, more experiments, and peer-to-peer problem solving.

For anyone championing a loonshot, anywhere:

- *Mind the False Fail:* See chapter 2 for the False Fail of Friendster (social networks) and the False Fails of the statins (the spurious results in mice and in dogs). Is a negative outcome due to a flaw in the idea or the test? What would you have to believe for it to be a flaw in the test? How might you evaluate that hypothesis?

- *Listen to the Suck with Curiosity (LSC):* When you have poured your soul into a project, you will be tempted to argue with critics and dismiss whoever challenges you. You will improve your odds of success by setting aside those urges and investigating, with genuine curiosity, the underlying reasons why an investor declines, a partner walks, or a customer chooses a competitor. It's hard to hear no one likes your baby. It's even harder to keep asking why. (Chapter 2)

- *Adopt a system rather than an outcome mindset:* Everyone will make wrong turns in navigating the long, dark tunnel through which every loonshot travels. You will gain much more (and feel much better) by trying to understand the *process* by which you arrived at those decisions. How did you prepare? What influenced you? How might you improve your decision-making process? (Chapter 5)

- *Keep your eyes on spirit, relationships, and time (SRT)*: A final word below, which is not in the main text. It's an added thought for anyone who makes it this far in the book.

When championing a loonshot, it's easy to lose sight of what's important, of why you are doing what you are doing. A little obsession can be good. Too much can backfire.

What's helped me, on occasion, to pull back from the edge—to create a more sustainable and productive level of obsession—is stepping back to think on SRT: spirit, relationships, and time.

Spirit

Some people find meaning in serving a higher power. Others find it in serving their country. Still others find it in providing for their families, or spreading joy, or helping others live better, freer lives. Everyone has a mission or noble purpose. William Faulkner, for example, spoke of the noble purpose of the writer and the poet:

> I believe that man will not merely endure: he will prevail. He is immortal, not because he alone among creatures has an inexhaustible voice, but because he has a soul, a spirit capable of compassion and sacrifice and endurance.
>
> The poet's, the writer's, duty is to write about these things. It is his privilege to help man endure by lifting his heart, by reminding him of the courage and honor and hope and pride and compassion and pity and sacrifice which have been the glory of his past.

When diving deep into a project or career it's easy for the head and the heart to stray to things that don't matter. I began in the academic world, in which the noble purpose is to seek truth. I switched to the biotech world, with a mission to improve the lives of patients in need. Both worlds, like all pursuits, offer fool's gold and true gold. Only by coming back to noble purpose could I tell the two apart.

Purpose feeds spirit, and spirit is the engine that keeps us going. It steadies us for the battles ahead.

Relationships

The support needed to survive the long tunnel of skepticism and uncertainty doesn't come from things. It comes from people. Several years ago, a physician who treats the terminally ill shared an insight with me that had changed his life. In hundreds of end-of-life conversations, he said, he never once heard anyone speak about what kind of car they have in their driveway, or even what kind of driveway they have. They always spoke of family and loved ones.

At the edge of obsession, relationships are often the first to go. But they are usually our most important need. When I catch myself making that mistake, I think back to those end-of-life conversations.

Time

The anxiety from championing a crazy idea, challenging experts, and facing repeated rejection can spill over into mindlessly filling a calendar. Completing urgent, but not important, tasks creates a sense of accomplishment and control. But time is our most precious resource, just as relationships are our most precious source of joy and support.

We all juggle many balls, a wise friend named Philip Lader likes to say, but what makes all the difference is knowing which are made of rubber and which are made of glass. For me, the ones to handle with great care, to avoid dropping and losing forever, have always been spirit, relationships, and time.

APPENDIX B
THE INNOVATION EQUATION

The simplified model company described in chapter 7 (physicists would call it a toy model) is useful for illustrating the idea of a phase transition inside organizations and deriving the equation shown in the text.

The organization design is the tree shown in the figure in chapter 7 in the tug-of-war section. The management span S is constant. At the bottom level of the tree (level $\ell = 1$) are the "doers," employees who do the work on projects (associates at a client service firm, programmers at a software firm, etc.). The next level up ($\ell = 2$) is the layer of project managers. The third level up and higher are the layers of professional managers (managers of managers: regional supervisors, functional supervisors, etc.). The total number of employees at the company is then $N = (S^L - 1)/(S - 1)$ where L is the total number of levels. When $L = 2$, for example, there are two levels, which is just one project, and the number of employees is $S + 1$.

Write each employee's compensation as $C = C_S + C_E$, where the first component is base salary, the second is equity stake in the form of equity units—for example, restricted stock or stock options. (It's not hard to add an annual *cash* bonus term, but let's start with the simplest model.) Assume salary increases by a constant fraction g with level, so base salary $C_S = C_{S0}\,(1 + g)^{\ell-1}$. Write the value of the equity stake as a percentage of face value: $C_E = a\,N_{EU}(\ell)P_{sh}$, where N_{EU} is the number of equity units owned on average at the level ℓ; and P_{sh} is the company's share price, and the constant a (which will not matter for the conclusions) is the same percentage for all employees. Approximating the value of equity stakes as a simple multiple of face value is often used in practice inside companies, when finer details from more sophisticated models, such as Black-Scholes, are not essential. (The widely used Radford employee compensation survey, as well as public company reporting guidelines, both follow this practice.) Next, assume that equity ownership increases by the same fraction as salary, so $N_{EU}(\ell) = N_{E0}(1 + g)^{\ell-1}$. In other words, when $g = 15\%$, each employee's base salary and equity stake increase by 15 percent on each promotion.

To connect employees' efforts and company value, assume that the enterprise value of the organization V_{ent} is the sum of the expected value of individual projects (i.e., nonproject contributions are small), and assume, for simplicity, each project has the same expected value V_0. To connect behavior and incentives, write x as the fraction of work time employees at level ℓ choose to spend on maximizing the expected value of projects in their span of control (V_{sp}), and y as the fraction spent on politicking (maximizing the likelihood of promotion to the next level of management, separate from project work). Assume total work time is fixed, so $x + y = 1$.

It is not hard to show that for an employee at level ℓ with span of projects V_{sp}, the change in compensation incentive from increasing the fraction of time y spent on politics is

$$\frac{d\ln C}{dy} = \tilde{g}\, R_P - \left(\frac{C_E}{C}\right)\left(\frac{V_{sp}}{V_{ent}}\right) R_S.$$

Here $\tilde{g} = \ln(1 + g)$; the *return on politics*, $R_p = (d\ell/dy)$, is the likelihood that incremental politicking will yield a promotion; and the *return on technical skill* $R_s = (d \ln V_{sp}/dx)$, is the percentage improvement in the value of assigned projects as a result of an incremental increase in time devoted to work (called *project–skill fit* in the text). In this model, a company with employees fully focused on their projects ($x = 1$, $y = 0$), will remain free of politics as long as the quantity above remains negative. In other words, when the politics term (first term on the right side) is smaller than the project term (second term on the right side), so that increasing time on politics (y) decreases compensation incentive. We can call the resulting inequality the *no-politics condition.*

Within the professional manager group, since span increases with rank, politics will appear first in the lowest layer ($\ell = 3$). Each of those managers spans S projects. The total number of projects in the organization is $S^{L-2} \approx N/S$ (for large S). Therefore $(V_{sp}/V_{ent}) = S^2/N$ for this layer. The inequality that ensures no politics among managers in the organization therefore becomes:

$$N < \frac{ES^2F}{\tilde{g}}.$$

Here the equity fraction $E = C_E/C$; organizational fitness (as in the text) $F = R_S/R_P$; and $\tilde{g} = \ln(1+g) \approx g$ (for small g). This gives the equation in the text.

* * *

Experts will recognize that the idea of a phase transition is defined, in the strict mathematical sense, only in the thermodynamic limit of an infinite number of interacting bodies. Yet scientists frequently apply the ideas to non-infinite systems: cars on a highway, for example—or humans at rock concerts.

The May 2013 issue of *Physical Review Letters*, the field's most prestigious rapid-communications journal, featured an article analyzing phase transitions at heavy metal concerts. Analysis revealed "a disordered gas-like state called a *mosh pit* and an ordered vortex-like state called a *circle pit.*" The authors identified phase separation at those concerts and provided a two-parameter phase diagram to help readers visualize the dynamics (as we did in with forests and highways in chapter 6).*

Although it may sometimes feel that way, the number of humans at

* The data sample included over one hundred YouTube concert videos watched by the authors, from which they concluded mosh pits and circle pits "are robust, reproducible, and largely independent of factors such as the musical subgenre, timing of performance, crowd size, arena size, suggestions from the band, time of year, and socioeconomic status of the moshers" (Silverberg et al., "Collective Motion of Humans in Mosh and Circle Pits at Heavy Metal Concerts," *PRL* 110 [2013]).

heavy metal concerts is not actually infinite. The science of phase transitions is used to describe finite systems because it helps us understand how and why large systems suddenly change. We use that understanding to help us manage those systems: as a source of ideas for designing safer forests, better highways, or—as we are doing here for the first time—more innovative companies.

The small $1/N$ effects of finite rather than infinite N don't matter much for those purposes (for systems with more than a hundred components, the finite-size effects will tend to be smaller than 1 percent).

Purists, however, may want to see that some kind of large-N limit can be defined in which there is a phase transition in the usual strict sense. That is the case in the model on the previous page. We can consider a compensation step-up rate that decreases as N increases, so that $\tilde{g}N \rightarrow g_0$, where g_0 is a constant independent of N. In that case, the system will undergo a transition, in the large-N limit, at a critical management span (steepness of the organizational tree) defined by $S^2_{\text{critical}} = g_0/EF$. A different but related model, without the assumption of a vanishing compensation step-up rate, also exhibits a well-defined transition in the large-N limit. If we assume a constant *cash* bonus fraction (as opposed to equity), we find a similar no-politics condition as above, but with ES^2/N replaced by the bonus fraction B. That model also exhibits a well-defined transition in the large-N limit.

ILLUSTRATION CREDITS

7 Scientist and piranha: Antar Dayal, AntarWorks LLC.

11 Water, ice, and hammer: Antar Dayal, AntarWorks LLC.

13 Tug-of-war: Antar Dayal, AntarWorks LLC.

25 Life on the edge: Antar Dayal, AntarWorks LLC.

30 The Battle of the Atlantic: Antar Dayal, AntarWorks LLC.

32 "Through superiority in the field of science": Antar Dayal, AntarWorks LLC.

37 Bush, Conant, Compton, and Loomis at UC Berkeley. © 2010 The Regents of the University of California, through the Lawrence Berkeley National Laboratory.

41 Ice to water: Antar Dayal, AntarWorks LLC.

53 The first patient to receive a statin: Akira Yamamoto.

55 Akira Endo: Antar Dayal, AntarWorks LLC.

64 Brown, Endo, and Goldstein: Akira Endo.

75 Lindbergh and Trippe: Pan American World Airways, Inc. Records, Special Collections, University of Miami Libraries, Coral Gables, Florida.

81 The *China Clipper*: Clyde Sunderland, collection of SFO Museum.

83 Chang letter to Trippe (1947): Trippe Family Archive, courtesy of the Pan Am Historical Foundation.

89 Pan Am launches the Jet Age: Antar Dayal, AntarWorks LLC.

103 Audience for 3D movie: J. R. Eyerman/The LIFE Picture Collection/Getty Images.

104 Meroë Morse: Polaroid Corporation Records, Baker Library, Harvard Business School.

105 Land and daughter: Polaroid Corporation Records, Baker Library, Harvard Business School.

107 Edwin Land unveils the first instant-print picture: *New York Times*, February 22 © 1947.

108 William Wegman Polaroid photography: © William Wegman.

108 Andy Warhol Polaroid photography: © Bill Ray.

108 *Life:* The SX-70: Co Rentmeester/The LIFE Picture Collection/Getty Images.

119 Land and Polavision machines: AP Photo/Bill Polo.

128 "And he came down from the hill with the tablet": Antar Dayal, AntarWorks LLC.

131 *Tubby the Tuba*: Antar Dayal, AntarWorks LLC.

139 Bond battles an evil monkey: Antar Dayal, AntarWorks LLC.

141 Garry Kasparov: Reuters/Jeff Christensen.

156 Sir Isaac Explain'd: Public domain. Note: the cover has been edited for readability.

161 Nagoya Dome: Tadaki et al., "Phase Transition in Traffic Jam Experiment," *New J. Phys.* 15 (2013).

167 Humphrey Bogart: Yousuf Karsh.

175 How do crickets harmonize?: Antar Dayal, AntarWorks LLC.

181 Sample content from an online terror cell: *Science* 352 (June 17, 2016): 1459. Reprinted with permission from AAAS.

182 Map of an online terror network: *Science* 352 (June 17, 2016): 1459. Reprinted with permission from AAAS.

183 Predicting when conflict will erupt: *Science* 352 (June 17, 2016): 1459. Reprinted with permission from AAAS.

185 Tug-of-war: Antar Dayal, AntarWorks LLC.

205 Launch of *Sputnik*: By permission of the Marcus family. Library of Congress, Prints & Photographs Division, LC-DIG-ds-04944.

212 The DARPA team prepares: DARPA.

219 Shredding the Dead Sea Scrolls: Antar Dayal, AntarWorks LLC.

229 Richard Feynman: Tamiko Thiel, via Wikimedia Commons.

238 Einstein and Kepler: Antar Dayal, AntarWorks LLC.

250 Shen Kuo: Antar Dayal, AntarWorks LLC.

255 Hooke, Boyle, and air pump: Rita Greer, via Wikimedia Commons.

256 Papin's discovery: Public domain.

SOURCE NOTES

Because sources mostly do not overlap between chapters, a bibliography is provided for each chapter, to make it easier for the reader to browse related subjects.

INTRODUCTION

Miller's piranha: Company reports; interview with Richard Miller. *Amgen*: See Introduction note for *"unable to repeat"* on page 315. *Nokia*: Cord, 19, 39, 118–20; Fox; Baker; O'Brien; Troianovski. *Merck*: Company reports. For more on Merck and the statins, see chapter 2. **Quotations:** *"law firm with a drug"*: Goozner, 128; conversations with industry insiders. *"synonymous with success"*: Baker. *"least hierarchical"* and *"have a bit of fun"*: Fox. *"After* The Lion King*"*: Catmull, 130–131. *"more is different"*: Anderson. For good surveys of the field of emergence see Ball; Gell-Mann; Johnson; Laughlin; and Miller.

Anderson, P. W. "More Is Different." *Science* 177 (1972): 393.

Baker, Stephen, et al. "Nokia: Can CEO Ollila Keep the Cellular Superstar Flying High?" *Bus. Week* 3590 (1998): 54.

Ball, Philip. *Why Society Is a Complex Matter.* Springer, 2012.

Binder, Gordon M., and Philip Bashe. *Science Lessons.* Harvard, 2008.

Catmull, Edwin, and Amy Wallace. *Creativity, Inc.* Random House, 2014.

Cord, David J. *The Decline and Fall of Nokia*. Schildts & Söderströms, 2014.

Fox, Justin. "Nokia's Secret Code." *Fortune* 141 (2000): 160.

Gell-Mann, Murray. *The Quark and the Jaguar.* Macmillan, 1994.

Goldwasser, E. "Erythropoietin: A Somewhat Personal History." *Persp. Biol. Med.* 40 (1996): 18.

Goozner, Merrill. "The Longest Search: How Eugene Goldwasser and Epo Gave Birth to Biotech." *Pharm. Exec.* 24 (2004): 112.
Johnson, Steven. *Emergence.* Scribner, 2001.
Laughlin, Robert B. *A Different Universe.* Basic Books, 2005.
Miller, John H. *A Crude Look at the Whole.* Basic Books, 2016.
O'Brien, Kevin J. "Nokia's New Chief Faces a Culture of Complacency." *NY Times*, Sept. 26, 2010.
Troianovski, Anton, and Sven Grundberg. "Nokia's Bad Call on Smartphones." *WSJ*, July 19, 2012.
Welte, Karl. "Discovery of G-CSF and Early Clinical Studies." In *Twenty Years of G-CSF*, edited by G. Molineux et al., 15. Springer, 2012.

CHAPTER 1

Dr. Angelina Callahan of the History Office of the Naval Research Laboratory kindly provided transcripts from a 1953 interview with Leo Young.

For thorough accounts by military historians of the role of radar in World War II, see Allison, *New Eye*; Brown, *Radar History*; Burns, "Technology"; and Guerlac, "Radar War." Roberts's *Storm of War* offers an excellent, broad World War II history. Dimbleby's *Battle of the Atlantic* does the same for the theater described in this chapter. Two concise guides to the vast literature on World War II are in Weinberg, *Short Introduction*, and D. Kennedy, *Freedom from Fear.*

Introduction and **The *Dorchester*:** *Radar at the NRL:* Allison, 39–82; Christman, 43–56 (which has been overlooked by most histories); Guerlac, *WWII*, 42–92; Page, 19–40; Young interview. *Radar at Pearl Harbor:* Prange, 366–75; US Congress, "Pearl Harbor," part 27, 527–29. **Quotations:** *"secret war . . . in vain"*: Churchill, *Finest Hour*, 337. *"irrespective of fog"*: Allison, 40. *"since the airplane"*: US Joint Board on SIP, 1. *"well exceed"*: Allison, 116. *"wild dream"* and *"pained me"*: Christman, 49, 56.

How not to fight a war: Bush, *Action*; Wiesner; Zachary. **Quotations:** *"how not to fight"*: Bush, *Action*, 74. *"lower caste"*: Bush, *Arms*, 19. *"already had a chemist"*: Greenberg, 58. *"rifle and bayonet"*: US War Department, 474. *"damn professors"*: Kevles, 133. *"tendency to fight a war"*: Bush, *Action*, 89. *"almost with awe"*: *Time*, "Yankee." *"await achievement"*: Stewart, epigraph.

Gathering storm: Wiesner; Zachary, 61–117; Sherwood, 154. **Quotations:** *"sea captains"*: Zachary, 23. *"wing of the President"*: Kevles, 301. *"do-gooders"* and *"minor miracles"*: Bush, *Action*, 35. *"OK, FDR"*: Zachary, 112; Bush, *Action*, 36. *"explore the bizarre"* and *"end run"*: Bush, *Action*, 102 and 31–2.

Life at 32 Fahrenheit: *Military, academic resistance:* Bush, *Action*, 69–120; Kevles, 243–47, 254–58. *Loomis:* Alvarez, 309–31; Conant, 108–32, 213–17, 309. *Radar in England:* Phelps, 23–78; Rowe; Jones, *Secret War. Battle of Britain:* Bungay, 104, 122, 199, 334–36; Moore, 104; Murray, 47. *British-American relations:* Phelps, 124–25, 140. *Loomis* *and the Tizard Mission:* Bowen, 150–63; Conant, 179–208; Phelps, 152–62, 176–203, 214–45. **Quotations:** "*tight organization,*" "*close collaboration,*" "*damned civilian,*" and "*Army did not want*": Bush, *Action*, 28–30, 103, 104. "*describe that weapon*": Lovell, 15. "*finest hour*": Churchill, 1940. "*would be nil*": Bungay, 334. "*foul specimen*": D. Kennedy, 443.

Massacre: *Role of technology in the U-boat war:* Baxter, 37–52, 136–86; Brown, 334–48; Burns; Guerlac, "Radar War"; P. Kennedy, 5–73; Kevles, 302–15; US Joint Board on SIP; Williamson. *American coast:* Dimbleby, 243–65; Gannon, *Drumbeat*, 214–41. *U-boat blockade:* Bungay, 339; Churchill, *Ring*, 4–5; Dimbleby, 121–46, 327–43, 376–89; D. Kennedy, 566–72; P. Kennedy, 30–37, 66–116. In mid-December 1942, Britain's fuel supply was 300,000 tons for commercial use and one million tons for emergency military use. Consumption ran at 130,000 tons per month—a runway of 10 months (Roskill, 217–18). *LORAN:* See chapter 1 note for "*a pilot could calculate his location*" on page 318. **Quotations:** "*help yourselves!*" and "*tanker*": Dimbleby, 260, 246. "*mushroom cloud*": Gannon, *Drumbeat*, 216. "*massacre*": Churchill, *Hinge*, 96. "*swallowed*" and "*swept through*": Middlebrook, 250, 252.

One at a time: *May 1943 battles:* Burns, 343, 353–54; Churchill, *Ring*, 8–10; Dimbleby, 401–23; Gannon, *Black May*; Roberts, 370–71; Rohwer, 318–31. *Proximity fuse (spelled "fuze" at the time):* Baldwin; Baxter, 221–42; Bush, *Action*, 106–12; Conant, 271–72; Hartcup, 39–45. *OSRD achievements:* Bush, *Frontier*; Baxter. *Bush and the bomb:* Goldberg; Bush, *Action*, 56–63; Zachary, 189–217. **Quotations:** "*driven under water*" and *other U-boat messages:* Syrett, 130–34. "*one at a time*": Gretton, 156–57. "*some months past*": Baxter, 46. "*we had lost*": Doenitz, *Memoirs*, 341. "*funny fuse*": Baldwin, 279.

Endless Frontier and **Eight Nobel Prizes:** *FDR's hidden illnesses:* Goldsmith; Lomazow. *Vail, AT&T:* Gabel, 345–48; Galambos; Gertner, 20–28; Hoddeson; Paine, 221–44; Reich, *American*, 151–84; Reich, "Bell Labs," 512–18. *Jewett and Bush:* Bush, *Action*, 32–37; Jewett. **Quotations:** "*flat on its face*": Kenny. "*pacemaker*": Bush, *Frontier*, 19. "*the scientific method*": *NY Times*, "Defense." "*Soviet Russia*": Kaempffert. "*epoch-making*": *BusinessWeek*. "*a week to get to you*": *NY Times*, "Phone."

The Bush-Vail Rules: *Bond:* See chapter 5. *3M:* Hindo. *Jobs:* See chapter 5. **Quotations:** "*In embryo*": Bush, *Action*, 72. "*combat regiment*": Bush, *Action*, 26. "*task allocation*": Reich, "Bell Labs," 519. "*unbalancing the whole*": Vail, 351. "*military men*": Bush, *Action*, 298. "*child psychologist*" and "*what the devil*": Bush, *Action*, 40, 111. "*I've seen*": Kevles, 309. "*hot potato*": Bush, *Action*, 45–46.

Allison, David K. *New Eye for the Navy*. Naval Research Lab, 1981.
Alvarez, Luis. "Alfred Lee Loomis." In *Biog. Mem. vol. 51*. Nat. Acad., 1980.
Baldwin, Ralph B. *The Deadly Fuze*. Presidio, 1980.
Baxter, James Phinney. *Scientists Against Time*. Little Brown, 1946.
Bernstein, Barton J. "American Conservatives Are the Forgotten Critics of the Atomic Bombing of Japan." *Merc. News*, July 31, 2014.

Bowen, E. G. *Radar Days*. Adam Hilger, 1987.

Breslin, Jimmy. *Branch Rickey*. Penguin, 2011.

Brown, Louis. *A Radar History of World War II*. Institute of Physics, 1999.

Budiansky, Stephen. *Blackett's War*. Knopf, 2013.

Bungay, Stephen. *The Most Dangerous Enemy: A History of the Battle of Britain*. Aurum, 2000.

Burns, R. W. "Impact of Technology on the Defeat of the U-Boat September 1939–May 1943." *IEE Proc. Sci. Meas. Tech.* 141 (1994): 343.

Bush, Vannevar. *Science: The Endless Frontier*. 1945.

———. *Modern Arms and Free Men*. Simon and Schuster, 1949.

———. *Pieces of the Action*. Morrow, 1970.

Business Week. "The Trend: 'Science—The Endless Frontier.'" July 21, 1945.

Christman, Albert B. *Target Hiroshima*. Naval Institute, 1998.

Churchill, Winston. *The Second World War, Vol. II: Their Finest Hour*, 1986 [1949]; *Vol. IV: The Hinge of Fate*, 1986 [1950]; *Vol. V: Closing the Ring*, 1986 [1951]. Houghton Mifflin Harcourt.

———. "Speech in the House of Commons." July 30, 1934; June 18, 1940.

Conant, Jennet. *Tuxedo Park*. Simon and Schuster, 2002.

CSPO. "Science the Endless Frontier 1945–1995." Center for Science, Policy and Outcomes, Columbia, 1998.

Dimbleby, Jonathan. *The Battle of the Atlantic*. Viking, 2015.

Doenitz, Karl. *Memoirs: Ten Years and Twenty Days*. World, 1959.

Einstein, Albert. Letter to Franklin D. Roosevelt, dated Aug. 2, 1939; delivered Oct. 11, 1939.

England, J. Merton. *A Patron for Pure Science*. NSF, 1983.

Erskine, Ralph. "Tunny Reveals B-Dienst Successes against the 'Convoy Code.'" *Intel. Nat. Sec.* 28 (2013): 868.

Fisher, David E. *A Race on the Edge of Time*. McGraw-Hill, 1987.

———. *A Summer Bright and Terrible*. Shoemaker & Hoard, 2005.

Gabel, Richard. "The Early Competitive Era in Telephone Communication, 1893–1920." *Law Cont. Prob.* 34 (1969): 340.

Galambos, Louis. "Theodore N. Vail and the Role of Innovation in the Modern Bell System." *Bus. Hist. Rev.* 66 (1992): 95.

Gannon, Michael. *Operation Drumbeat*. Harper & Row, 1990.

———. *Black May*. HarperCollins, 1998.

Gardner, W. J. R. *Decoding History*. Macmillan, 1999.

Gertner, Jon. *The Idea Factory*. Penguin, 2012.

Goldberg, Stanley. "Inventing a Climate of Opinion: Vannevar Bush and the Decision to Build the Bomb." *Isis* 83 (1992): 429.

Goldsmith, Harry S. *A Conspiracy of Silence*. iUniverse, 2007.

Greenberg, Daniel S. *The Politics of Pure Science*. 1967. U. Chicago, 1999.

Gretton, Peter. *Convoy Escort Commander*. Cassel, 1964.

Guerlac, Henry. *Radar in World War II*, vol. 8. 1987. Am. Inst. Phys., 1946.

Guerlac, Henry, and Marie Boas. "The Radar War against the U-Boat." *Military Affairs* 14 (1950): 99.
Hartcup, Gordon. *The Effect of Science on the Second World War.* St. Martin's Press, 2000.
Hasegawa, Tsuyoshi. *Racing the Enemy.* Harvard, 2005.
———, ed. *The End of the Pacific War.* Stanford, 2007.
Hindo, Brian. "At 3M, A Struggle between Efficiency and Creativity." *Bloomberg.com*, June 11, 2007.
Hoddeson, Lillian. "The Emergence of Basic Research in the Bell Telephone System, 1875–1915." *Tech. Cult.* 22 (1981): 512.
Jarboe, Kenan, and Robert Atkinson. "The Case for Technology in the Knowledge Economy." Prog. Pol. Inst., 1998.
Jewett, F. B. "The 1943 Medalist." *Elec. Eng.* 63 (1944): 81.
Jones, Reginald V. "Winston Leonard Spencer Churchill, 1874–1965." *Biog. Mem. Fell. Roy. Soc.* 12 (1966): 35.
———. *Most Secret War.* Penguin, 2009.
Kaempffert, Waldemar. "Dr. Bush Outlines a Plan." *NY Times*, July 22, 1945.
Kelly, Cynthia C., ed. *Manhattan Project.* Black Dog & Leventhal, 2007.
Kennedy, David M. *Freedom from Fear.* Oxford, 1999.
Kennedy, Paul M. *Engineers of Victory.* Random House, 2013.
Kenny, Herbert. "At 80, Scientist Bush Looks Back at Eventful Years." *Boston Globe*, Sep. 20, 1970.
Kevles, Daniel J. *The Physicists.* Knopf, 1971.
Lane, Julia. "Assessing the Impact of Science Funding." *Science* 324 (2009): 1273.
Lomazow, Steven, and Eric Fettmann. *FDR's Deadly Secret.* PublicAffairs, 2009.
Lovell, Stanley P. *Of Spies & Stratagems.* Pocket Books, 1963.
Middlebrook, Martin. *Convoy SC.122 & HX.229.* Allen Lane, 1976.
Moore, Kate, and the Imperial War Museum. *The Battle of Britain.* Bloomsbury, 2010.
Murray, Williamson. *Strategy for Defeat: The Luftwaffe, 1933–1945.* Air Univ. Press, 1983.
NAS. *Rising above the Gathering Storm, Revisited.* Nat. Acad. Press, 2010.
New York Times. "Phone to Pacific from the Atlantic." Jan. 26, 1915.
———. "Tesla, At 78, Bares New 'Death Beam.'" July 11, 1934.
———. "Research for Defense." July 21, 1945.
Page, Robert M. *The Origin of Radar.* Anchor Books, 1962.
Paine, Albert Bigelow. *In One Man's Life: Being Chapters from the Personal & Business Career of Theodore N. Vail.* Harper, 1921.
PCAST (President's Council of Advisors on Science & Technology). "Report to the President: Transformation and Opportunity; The Future of the U.S. Research Enterprise," 2012.
Perutz, M. F. "That Was the War." *New Yorker*, Aug. 12, 1985.
Phelps, Stephen. *The Tizard Mission.* Westholme, 2010.
Prange, Gordon W., et al. *Pearl Harbor.* McGraw-Hill, 1986.

Reich, Leonard S. "Industrial Research and the Pursuit of Corporate Security: The Early Years of Bell Labs." *Bus. Hist. Rev.* 54 (1980): 504.
———. *The Making of American Industrial Research.* Cambridge, 1985.
Rinzler, J. W. *The Making of* Star Wars. Del Rey, 2007.
Roberts, Andrew. *The Storm of War.* HarperCollins, 2009.
Roosevelt, Franklin D. Cable communication with Winston Churchill, May 4, 1941.
———. Letter to Vannevar Bush, Nov. 17, 1944.
Roskill, Stephen W. *The War at Sea, 1939–1945: Vol. II.* HMSO, 1954.
Rohwer, Jürgen. *Chronology of the War at Sea, 1939–1945. Vol. II: 1943–1945.* Trans. D. Masters. Ian Allen, 1974.
Rowe, Albert P. *One Story of Radar.* Cambridge, 1948.
Sherwood, Robert E. *Roosevelt and Hopkins*, revised ed. Harper, 1950.
Snow, C. P. *Science and Government.* Harvard, 1961.
Stewart, Irvin. *Organizing Scientific Research for War.* Little, Brown, 1948.
Syrett, David. *The Defeat of the German U-Boats.* U. South Carolina, 1994.
Tighe, W. G. S. "Review of Security of Naval Codes and Ciphers, September 1939 to May 1945." Public Record Office, 1945.
Time. "Yankee Scientist." Apr. 3, 1944.
US Congress Joint Committee on the Investigation of the Pearl Harbor Attack. *Hearings before the Joint Committee.* USGPO, 1946.
US Joint Board on Scientific Information Policy. *Radar: A Report on Science at War.* 1945.
US War Department, *Annual Reports.* USGPO, 1920.
Vail, Theodore N. *Views on Public Questions; a Collection of Papers and Addresses of Theodore Newton Vail, 1907–1917.* Priv. print., 1917.
Walker, J. Samuel. "Recent Literature on Truman's Atomic Bomb Decision: A Search for Middle Ground." *Dipl. Hist.* 29 (2005): 311.
———. Interview with Cindy Kelly, Mar. 16, 2014.
———. *Prompt and Utter Destruction*, third ed. UNC, 2016.
Weinberg, Gerhard L. *World War II: A Very Short Introduction.* Oxford, 2014.
Wiesner, Jerome. "Vannevar Bush." In *Biog. Mem.*, vol. 50. Nat. Acad., 1979.
Williamson, Gordon. *U-Boat Tactics in World War II.* Osprey, 2010.
Young, Leo. Interview. October 15, 1953.
Zachary, G. Pascal. *Endless Frontier.* MIT, 1999.

CHAPTER 2

The Three Deaths: *Gleevec:* Vasella; Wapner; Monmaney. *Tehran:* Bohlen, 141–45. *FDR's health:* Lomazow, 56–60, 94–101; Bruenn; Lahey; Winik, 485–86, 507–12, 520–21. *Early heart disease history:* Leibowitz, 1–103; Steinberg, 4, 9–10. *NHI, Framingham:* Mahmood; Steinberg, 36–37; Strickland, 52–53; Truswell, 21–24. *Mortality trends:* Ford; Jones; NHLBI, 23 (75%); NIH; interview with Paul Sorlie, NHLBI. **Quotations:** *"over my dead body"*: Wapner, 156. *"great drops of sweat"*: Bohlen, 143–44. *"out of the blue"*: Lomazow, 9. *"hardly had a place . . . quiet, liquor, opium"*: Heberden, 366–69.

Fungi don't run: *Cholesterol-lowering trials:* Steinberg, 125–74; Truswell, 51–62, 109–20. *Endo, first statin:* interviews with Akira Endo, Joseph Goldstein, and Ed Scolnick; Brown, "Tribute"; Daida; Endo, "Gift"; Endo, "Origin"; Endo, "Perspective." *Keys:* Keys; Steinberg, 33–35; *Time*; Taubes, 16–41; Truswell, 33–38; **Quotations:** *"disgusting"* and *"start to think"*: *Time*. *"like sumo wrestlers"*: Daida, 686.

Saved by chickens: *Cholesterol, Endo:* as above; Yamamoto; Steinberg, 174–86; Tobert. *Brown and Goldstein history:* Brown, "Side Trips"; Brown, "Nobel Lecture"; interview with Joseph Goldstein. **Quotations:** *"thousands of fat chemists"*: Mann, 645. *"All well-controlled trials"*: McMichael, 173. *"little more than zero"*: *Lancet*, 605. *"Gilbert and Sullivan"*: Steinberg, 104.

A $90 billion "coincidence": *Merck and Sankyo:* See chapter 2 note for *"to collaborate, rather than compete"* on page 321. *Within days of:* Li, 60; Steinberg, 178; Vagelos, 134. *Statin development and usage, reviews:* Collins; Goldfine; Goldstein; Li; Steinberg, 171–211; Truswell, 115–19; and references therein. **Quotations:** *"sudden, unbelievable, the thrill of discovery"*: Vagelos, 134–35. *"dramatic effect"*: Goldfine, 1752. *"fungal extracts"*: Brown, "Tribute," 16.

Counting the arrows in your ass: Begley; Cooke, *Folkman's War*, 180–87, 283–90, 296–99; Ferrara et al.; Folkman; Genentech; Rosenfeld; Stone. **Quotations:** *"go to the bathroom"* and *"cured cancer again"*: Begley. *"Spousal Activation Factor"* and *"clown"*: Folkman, "Fine Line," 4, 13. *"Maybe we don't have to die"*: McAlary. *"Judah will cure cancer"*: Kolata. *"sign for me"*: Folkman, *Acad. Ach.* *"careers wouldn't be harmed"*: Ezzell. *"only wish"*: Cooke, "Progeny." *"arrows in his ass"*: Cooke, *Folkman's War*, 154.

Lessons from the surprising fragility: *Social networks:* interview with Ken Howery; Cowley; Kirkpatrick, 87–90; Thiel. *Deak Parsons:* Christman, 42–56. *Folkman:* King; Cooke, *Folkman's*, 296–98. **Quotations:** *"little-known bit," "revolutionize naval weaponry,"* and *"door-to-door salesman"*: Christman, 43, 47, 49.

Alberts, Alfred W. "Discovery, Biochemistry and Biology of Lovastatin." *Am. J. Cardio.* 62 (1988): J10.

Begley, Sharon, et al. "One Man's Quest to Cure Cancer." *Newsweek*, May 18, 1998.

Bohlen, Charles E. *Witness to History, 1929–1969.* Norton, 1973.

Brown, Michael S., and Joseph L. Goldstein. "A Receptor-Mediated Pathway for Cholesterol Homeostasis." In *Nobel Lectures Phys. Med. 1981–1990*, 284. World Scientific, 1993.

———. "A Tribute to Akira Endo." *Athero. Supp.* (2004): 13.

———. "Scientific Side Trips." *J. Biol. Chem.* 287 (2012): 22418.

Bruenn, H. G. "Clinical Notes on the Illness and Death of President Franklin D. Roosevelt." *Ann. Int. Med.* 72 (1970): 579.

Christman, Albert B. *Target Hiroshima.* Naval Institute, 1998.

Collins, Rory, et al. "Interpretation of the Evidence for the Efficacy and Safety of Statin Therapy." *The Lancet* 388 (2016): 2532.

Cooke, Robert. *Dr. Folkman's War.* Random House, 2001.

———. “Dr. Folkman’s Progeny.” *Vector,* Spring 2008.

Cordes, Eugene H. *Hallelujah Moments.* Oxford, 2014.

Cowley, Stacy, and Julianne Pepitone. “Facebook’s First Big Investor, Peter Thiel, Cashes Out.” *CNNMoney,* Aug. 20, 2012.

Daida, Hiroyuki. “Meet the History: The Discovery and Development of ‘Statin,’ the Penicillin of Arteriosclerosis.” *Shinzo (Heart)* 37 (2005): 681.

Endo, Akira. “The Origin of the Statins.” *Athero. Supp.* 5 (2004): 125.

———. “A Gift from Nature: The Birth of the Statins.” *Nat. Med.* 14 (2008): 1050.

———. “A Historical Perspective on the Discovery of Statins.” *Proc. Jap. Acad. Series B* 86 (2010): 484.

Ezzell, Carol. “Starving Tumors of Their Lifeblood.” *Sci. Am.* 279 (1998): 33.

Ferrara, Napoleone, and Anthony P. Adamis. “Ten Years of Anti-Vascular Endothelial Growth Factor Therapy.” *Nat. Rev. Drug Disc.* 15 (2016): 385.

Ferrara, Napoleone, et al. “Discovery and Development of Bevacizumab.” *Nat. Rev. Drug Disc.* 3 (2004): 391.

Folkman, Judah. “Tumor Angiogenesis: Therapeutic Implications.” *New Eng. J. Med.* 285 (1971): 1182.

———. “The Fine Line between Persistence and Obstinacy in Research.” Speech given in 1996. Folkman collection, Harvard Medical Library, Countway Library of Medicine, Box 45 Folder 19.

———. Academy of Achievement Interview, June 18, 1999.

———. “Is Angiogenesis an Organizing Principle in Biology and Medicine?” *J. Ped. Surg.* 42 (2007): 1; and *Nat. Rev. Drug Disc.* 6 (2007): 273.

Ford, Earl S., et al. “Explaining the Decrease in U.S. Deaths from Coronary Disease, 1980–2000.” *New Eng. J. Med.* 356 (2007): 2388.

Genentech press release. “Positive Results from Phase III Avastin Study in Metastatic Colorectal Cancer,” June 1, 2003.

Goldfine, Allison B. “Statins: Is It Really Time to Reassess Benefits and Risks?” *New Eng. J. Med.* 366 (2012): 1752.

Goldstein, Joseph L., and Michael S. Brown. “A Century of Cholesterol and Coronaries.” *Cell* 161 (2015): 161.

Heberden, William M. D. *Commentaries on the History and Cure of Diseases.* 1806. Luke Hansard, 1772.

Jones, David S., and Jeremy A. Greene. “The Decline and Rise of Coronary Heart Disease.” *Am. J. Pub. Health* 103 (2013): 1207.

Kannel, William B., et al. “Factors of Risk in the Development of Coronary Heart Disease.” *Ann. Int. Med.* 55 (1961): 33.

Keys, Ancel, et al. “Lessons from Serum Cholesterol Studies in Japan, Hawaii and Los Angeles.” *Ann. Int. Med.* 48 (1958): 83.

King, Ralph T., Jr. “Human Test to Begin on Tumor Drug Despite Mixed Results of NCI Mice Tests.” *WSJ,* Sep. 13, 1999.

———. “Novel Cancer Approach from Noted Scientist Hits Stumbling Block.” *WSJ,* Nov. 12, 1998.

Kirkpatrick, David. *The Facebook Effect*. Simon & Schuster, 2010.
Kolata, Gina. "A Cautious Awe Greets Drugs That Eradicate Tumors in Mice." *NY Times*, May 3, 1998.
Lahey, Frank. "Memo Regarding the Health of Franklin Delano Roosevelt," July 10, 1944. The memo was publicly revealed only 63 years later, in Goldsmith, Harry S., *A Conspiracy of Silence*, iUniverse, 2007.
Lancet. "Can I Avoid a Heart-Attack?" 1 (1974): 605.
Leibowitz, Joshua O. *The History of Coronary Heart Disease*. Wellcome, 1970.
Li, Jie Jack. *Triumph of the Heart*. Oxford, 2009.
Lomazow, Steven, and Eric Fettmann. *FDR's Deadly Secret*. PublicAffairs, 2009.
Mahmood, Syed S., et al. "The Framingham Heart Study." *The Lancet* 383 (2014): 999.
Mann, George V. "Diet-Heart: End of an Era." *New Eng. J. Med.* 297 (1977): 644.
McAlary, Mike. "I Cling to This Hope for Life." *NY Daily News*, May 6, 1998.
McMichael, John. "Fats and Atheroma: An Inquest." *Br. Med. J.* 1 (1979): 173.
Monmaney, Terence. "The Triumph of Dr. Druker." *Smithsonian* 42 (2011): 54.
NHLBI: National Heart Lung and Blood Institute. "Morbidity & Mortality: 2012 Chart Book on Cardiovascular, Lung, and Blood Diseases."
NIH: National Institutes of Health. "Heart Disease Fact Sheet." Oct. 2010.
NRC: National Research Council. NRC/1980: *Issues and Current Studies*. Nat. Acad. Sci., 1981.
Rosenfeld, Philip J., et al. "Ranibizumab for Neovascular Age-Related Macular Degeneration." *New Eng. J. Med.* 355 (2006): 1419.
Steinberg, Daniel. *The Cholesterol Wars*. Academic Press, 2007.
Stone, Edwin M. "A Very Effective Treatment for Neovascular Macular Degeneration." *New Eng. J. Med.* 355 (2006): 1493.
Strickland, Stephen P. *Politics, Science, and Dread Disease*. Harvard, 1972.
Taubes, Gary. *Good Calories, Bad Calories*. Knopf, 2007.
Thiel, Peter. *The James Altucher Show*, Oct. 2, 2014.
Time. "Medicine: The Fat of the Land." Jan. 13, 1961.
Tobert, Jonathan A. "Lovastatin and Beyond." *Nat. Rev. Drug Disc.* 2 (2003): 517.
Truswell, A. Stewart. *Cholesterol and Beyond*. Springer, 2010.
Vagelos, P. Roy, and Louis Galambos. *Medicine, Science and Merck*. Cambridge, 2004.
Vasella, Daniel. *Magic Cancer Bullet*. HarperBusiness, 2003.
Wapner, Jessica. *The Philadelphia Chromosome*. The Experiment, 2013.
Winik, Jay. *1944: FDR and the Year That Changed History*. Simon & Schuster, 2015.
Woodruff, H. Boyd., Exec. Adm. Merck Sharp & Dohme Research Laboratories Division. Letters to Issei Iwai, Director of Product Planning Dept., Sankyo, Apr. 16, 1976, and Sep. 23, 1977.
Yamamoto, Akira, Hiroshi Sudo, and Akira Endo. "Therapeutic Effects of ML-236B in Primary Hypercholesterolemia." *Athero.* 35 (1980): 259.
Yancopoulos, George D. "Clinical Application of Therapies Targeting VEGF." *Cell* 143 (2010): 13.

CHAPTER 3

Doug Miller of the Pan Am Historical Foundation; Nicola Hellmann-McFarland of the University of Miami Libraries, Coral Gables, special collections department; and Ed Trippe kindly provided access to archival materials.

Introduction and **The two types:** *Pan Am brand:* Gandt, 54–55; Petzinger, 19; Verhovek, xix–xx; Davey. *Decline:* Dunlap; Lester, 263–69; Pyle. *Google, Walmart:* Doerr, 3–5; Vise, 40–43; Walton, 32–54. *Facebook:* see chapter 2. *Airline bankruptcies:* Cohen, C224; Dempsey, 427. *Eclipse of 1919:* Overbye. **Quotation:** *"just a toy"*: Paine, 98–99.

JT and Crando: *Trippe:* Bender, 22–26, 101–2, 135; Daley, 95–96. *Crando:* Petzinger, 350–51; Rubin; Zellner. *Long Island:* Bender, 60–63; Daley, 9–11. **Quotations:** *"Yale gangster"*: Bender, 13. *"politest"*: Bender, 490. *"if you don't win"* and *"Where were you?"*: Maxon. *"batshit"*: Petzinger, 55. *"academic eggheads"*: Petzinger, 102 ("Crandall does not recall the comment but does not deny making it"). *"kerosene in his blood"*: Rubin.

The pie industry: *AA innovations*: Cohen; Petzinger; Reed; Rubin. **Quotations:** *"legalized warfare"*: Rubin. *"cannibalistic"*: Zellner. *"Tuesday as opposed to Friday"* and *"yield management"*: Petzinger, 304.

JT and Lindy: *C. A. Lindbergh (CAL) flight and mania:* Berg, 112–32, 135–63; CAL, *AoV*, 70–83; Jackson, 271–77, 305–25; Van Vleck, 42–44. *Prior attempts:* Connor; Jackson, 369–70. *CAL, Trippe, Latin America:* Bender, 100–101; Berg, 172–75, 191; CAL, *AoV*, 83–96; Van Vleck, 56–64; Daley, 60–68, 484. *S-38:* Bender, 182–85, 100–101; Davies, 12–13. *Radio navigation, Fatt crash:* Aeronautical—1929, 108–9; Aeronautical—1930, 52; Bender, 155–63; Daley, 43–55. *Miles flown:* Daley, 484. *Honeymoon tour:* Bender, 136–40; Berg, 214–18; *NY Times*, "Lindbergh Log"; *NY Times*, "Four On Flight." **Quotations:** *"flight of the gulls"*: Daley, 41. *"load shifting," "Nordic race," "smiling American boy,"* and *"Not even Columbus"*: Berg, 108, 116, 121, 142. *"boy in evening dress"*: A. Lindbergh, 89. *"three hundred miles"*: CAL, *AoV*, 93–94. *"not enough to keep me busy"*: Daley, 55.

The dangerous virtuous cycle: *Atlantic:* Daley, 105–8; Davies, 30; Van Vleck, 93. *Pacific:* Bender, 226; Conant; Daley, 152–53; Davies, 36; Jackson, 369–70; Van Vleck, 92. *Wake, Midway, Guam:* Daley, 106–8, 135–44; Van Vleck, 95. *China Clipper:* Bender, 248–53; Daley, 165–75; Davies, 38; Nugent; Van Vleck, 101. **Quotations:** *"flying boat," "about to witness," "glorious history,"* and *"we all ducked"*: Daley, 110, 169–74.

Wars, loonshots, and cuckoo clocks: *FDR, Churchill:* Bender, 329–68; Daley, 302–14, 336–37; Van Vleck, 135–36. *Chang meeting:* Daley, 387–95, 512–13; Chang; Leslie. *Rockets:* see chapter 3 note for *"dismissed by academics and the military"* on page 323; Ford, 137; Brentford. *CAL vs. FDR:* Olson. *CAL, Goddard, Germany:* Berg, 210–14, 463–70; CAL, *Journals*, 101, 104–5, 955–59. *Return:* Daley, 330–32, 397–98; Gandt, 29–30. **Quotations:** *"high schools"* and *"regrets the error"*: *NY Times*, Jan. 13, 1920, and July 17, 1969. *"your own Dr. Goddard"*: Berg, 472. *"dented the myth"*: Arthur Schlesinger, cited in Olson, xv. *"Our next President"*: Berg, 419. *"appeaser"*: *NY Times*,

"Lindbergh Quits." *"young man's wings," "from Jesus to Judas"*: Berg, 437, 433. *"white and shaky"*: CAL, *Journals*, 958.

The Jet Age: *Comet:* Anderson, 202–4; Bender, 467–68; Pushkar; Verhovek, 7–9, 17–18. *The 707:* Bender, 469–76; Daley, 396–414; Verhovek, 28–29, 108–10; *WSJ*. *Hotels, missiles, moon:* Bender, 486; Daley, 426–27; Gandt, 8–9. **Quotations:** *"Elizabethan era"*: Verhovek, 10. *"earthquake victim"*: Bender, 475.

One more turn: *The 747:* Daley, 430–39; Bender, 500–507; Petzinger, 19. *Monopoly weakens:* Daley, 346, 420; Petzinger, 99–100. *Deregulation, S-type loonshots:* Petzinger, 31–32, 122–53, 286, 351–52. *Decline:* Bender, 515–25; Bennett; Crittenden; Daley, 440–49; Gandt, 283–84; Petzinger, 100–101. **Quotation:** *"If you buy it"*: Irving, 195.

Watch your blind side: *IBM, Microsoft, Intel:* Chandler, 118; Chposky, 45; McGregor; Sanger.

Aeronautical Chamber of Commerce: Aircraft Year Books 1929 and 1930.
Anderson, Dale, et al. *Flight and Motion*. Sharpe, 2009.
Bender, Marylin, and Selig Altschul. *The Chosen Instrument*. Simon and Schuster, 1982.
Bennett, Robert A. "Pan Am's Disappearing Act." *NY Times*, Jan. 18, 1987.
Berg, A. Scott. *Lindbergh*. Putnam, 1998.
Boyne, Walter J. *The Messerschmitt Me 262*. Smithsonian, 1980.
Brentford and Chiswick Local History Society. "Finding Private Browning," Sep. 23, 2004.
Bush, Vannevar. *Modern Arms and Free Men*. Simon and Schuster, 1949.
Chandler, Alfred Dupont, et al. *Inventing the Electronic Century*. Harvard, 2001.
Chang, Kia-ngau. Letter to Juan T. Trippe, Sep. 8, 1947, and "Memorandum Re China," n.d. 1947 (Trippe Family Archive, courtesy Pan Am Historical Foundation).
Christensen, Clayton M. *The Innovator's Dilemma*. Harvard, 1997.
Chposky, James, and Ted Leonsis. *Blue Magic*. Facts on File, 1988.
Clary, David A. *Rocket Man*. Hyperion, 2003.
Cohen, Isaac. "American Airlines," in *Strategic Management*, 10th ed., edited by C. Hill and G. Jones. Cengage Learning, 2013, C224.
Conant, Jane Eshleman. "Dole Air Race—1927." *SF Call-Bulletin*, Oct. 10, 1955.
Connor, Roger. "Even Lindbergh Got Lost." *Air & Space Mag.* Accessed Oct. 6, 2017.
Crittenden, Ann. "Juan Trippe's Pan Am." *NY Times*, July 3, 1977.
Daley, Robert. *An American Saga*. Random House, 1980.
Davey, Helen. "Orphaned by Job Loss." *Huffington Post*, April 4, 2010.
Davies, R. E. G. *Pan Am: An Airline and Its Aircraft*. Orion, 1987.
Dempsey, Paul. "The Financial Performance of the Airline Industry Post-Deregulation." *Houston Law Rev.* 45 (2008): 421.
Doerr, John. *Measure What Matters*. Penguin, 2018.

Dunlap, David W. "Final Pan Am Departure." *NY Times*, Sep. 4, 1992.

Ford, Brian J. *Secret Weapons*. Bloomsbury, 2011.

Galambos, Louis. "Theodore N. Vail and the Role of Innovation in the Modern Bell System." *Bus. Hist. Rev.* 66 (1992): 95.

Gandt, Robert. *Skygods*. William Morrow, 1995.

Irving, Clive. *Wide-Body: The Triumph of the 747*. William Morrow, 1993.

Jackson, Joe. *Atlantic Fever*. Farrar, Straus and Giroux, 2012.

King, Benjamin, and Timothy Kutta. *Impact: The History of Germany's V-Weapons in World War II*. Sarpedon, 1998.

Leslie, John C. Letter to Samuel F. Pryor, April 25, 1974. (Courtesy of Pan Am Records, Special Collections, U. of Miami Libraries, Coral Gables, FL.)

Lester, Valerie. *Fasten Your Seat Belts*. Paladwr, 1995.

Lindbergh, Anne M. *Bring Me a Unicorn: Diaries and Letters, 1922–1928*. HBJ, 1972.

Lindbergh, Charles A. *The Wartime Journals of Charles A. Lindbergh*. HBJ, 1970.

———. *Autobiography of Values*. HBJ, 1978.

MacDonald, Charles. "Lindbergh in Battle." *Collier's* 117 (1946).

Maxon, Terry. "Tales from the Beat: Robert L. Crandall." *Dallas News*, Sep. 6, 2015.

May, Ernest R. "1947–48: When Marshall Kept the US out of War in China." *J. Mil. His.* 66 (2002): 1001.

McGregor, Jena. "The Biggest Mass Layoffs of the Past Two Decades." *Wash. Post*, Jan. 28, 2015.

NY Times. "Lindberghs to Take Four on Flight South." Sep. 18, 1929.

———. "Lindbergh Log Sent by Radio Operator." Sep. 23, 1929.

———. "Lindbergh Quits Air Corps; Sees His Loyalty Questioned." April 29, 1941.

Nugent, Frank. "Warner's China Clipper." *NY Times*, Aug. 12, 1936.

Olson, Lynne. *Those Angry Days*. Random House, 2013.

Overbye, Dennis. "The Eclipse That Revealed the Universe." *NY Times*, July 31, 2017.

Paine, Albert Bigelow. *In One Man's Life: Being Chapters from the Personal and Business Career of Theodore N. Vail*. Harper, 1921.

Pavelec, Sterling Michael. *The Jet Race and the Second World War*. Praeger, 2007.

Petzinger, Thomas. *Hard Landing*. Times Business, 1995.

Pushkar, Robert. "Comet's Tale." *Smithsonian*, June 2002.

Pyle, Mark. "December 4, 1991: The Last 'Clipper' Flight." *Airways Mag.*, Dec. 4, 2016.

Reed, Dan. *The American Eagle*. St. Martin's Press, 1993.

Rubin, Dana. "Bob Crandall Flies Off the Handle." *Texas Monthly*, Aug. 1993.

Sanger, David E. "IBM Sells Back Much of Intel Stake." *NY Times*, June 12, 1987.

Van Vleck, Jenifer. *Empire of the Air*. Harvard, 2013.

Verhovek, Sam. *Jet Age*. Avery, 2010.

Vise, David A., and Mark Malseed. *The Google Story*. Bantam Dell, 2005.

Wall Street Journal. "Pan American Orders 25 Jet Planes." Oct. 14, 1955.

Wallace, Max. *The American Axis.* St. Martin's Press, 2004.
Walton, Sam. *Made in America.* Doubleday, 1992.
Wedemeyer, General Albert C. *Wedemeyer Reports!* Holt, 1958.
Zellner, Wendy, and Andrea Rothman. "The Airline Mess." *Bus. Week* 3273 (1992): 50.

CHAPTER 4

Introduction, Han Solo's escape, and **Disappearing fish:** *Polaroid achievements:* below. *Early Land:* Campbell, 198–200; Fierstein, 4–14; McElheny, 21–47. *Headlight glare:* Fierstein 5–6, 40–41; McElheny, 23, 28–29, 35–36, 50–57, 71, 86–107. *Sunglasses:* Campbell, 199–200; Fierstein, 20, 33–34; McElheny, 63–64; *Newsweek*, "General Patton." **Quotations:** *"stunning"*: *Time*, 81. *"remarkable"*: Cordtz, 85. *"can't stop using it"*: *Fortune*, 31. *"nearly impossible"*: Land, "Letter." *"read the Bible," "doing to excess," "obsolete words," "most exciting single event," "see into my head,"* McElheny, 20, 34, 37, 60, 420. *"After the bellboy . . . anything like this"*: Manchester, *New World*, 44. *"every second man"*: *Life*, "Light Control," 72.

From art fair to warfare: *Harvard:* C. Kennedy papers; McElheny, 109–11; Middeldorf; *NY Times*, "Kennedy." *Vectograph:* Fierstein, 31–37; McElheny, 115–18, 126–39. *3D movies:* McElheny, 118–25. *Adams, artists:* Adams, 293–307; Bonanos, 8–13; Fierstein, 24–25, 71–73. *Diversity:* Bonanos, 14–15, 35–36; *Life*, "Bonanza," 82; McElheny, 59–60, 210–18. **Quotations:** *"more beautiful than the originals"*: Kennedy, 212. *"irritate"*: Middeldorf, 373. *"hear the dripping water," "soul mate"*: McElheny, 19, 219.

An obvious question: *Inventing instant pictures:* Fierstein, 42–53; Laurence; McCune; McElheny, 164–88. *Lean times:* Fierstein, 52; Wensberg, 11, 90–91. *Countless advances:* overviews, Bonanos; more technical, McElheny. *Sales:* Bonanos, 55; Holt. *New molecule:* McElheny, 233. *Color vision:* Land, "Experiments"; McElheny, 245–77; F. Smith. *El Capitan, Carter:* Adams, 302–6; McElheny, 205. *Intimacy pictures:* Bonanos, 71–73. **Quotations:** *"see them now"* and *"from 1943 to 1972"*: McElheny, 163. *"meet me at five"* and *"red phone"*: Bonanos, 21. *"lots more outgo"* and *"Everyone went wild"*: McCune. *"greatest advances"*: Manchester, "60 Seconds," 167. *"what Mr. Land has done"*: *NY Times*, "Does the Rest." *"Nobel Prizes"*: Bernstein.

Polavision: *Instant movies:* Bonanos, 111–21; Czach; McElheny, 409–24. *Small successes and final failure:* Bonanos, 117; McElheny, 431–35. *Land's exit:* Bonanos, 135–56. **Quotations**: *"what the phonograph does for the Ear"*: Edison. *"a second revolution"*: Czach. *"impossible concepts into hardware"*: Ortner. *"the film I had just made"*: Fantel. *"highlight of his career"*: Shumacher. *"more scientific and aesthetic appeal"*: *Time*, "Instant Movies." *"cruel misuse of language"*: McElheny, 433.

Photons, electrons, and Richard Nixon: *CCDs:* Boyle; G. Smith. *Polaroid's digital efforts:* Tripsas. *NRO declassified materials:* Perry. *The Land Panel, U2:* Burrows, 110–27; McElheny, 278–305, 322–40; Norris, 49–51; Richelson, 12–15, 51–54, 58–59; TCP; Wang,

49–53. *Czechoslovakia:* Richelson, 170. *The battle over CCD surveillance:* Perry, 521–33. *KH-11:* Burrows, 217–21; Perry, 77–81; Richelson, 198–202. **Quotations:** *"second term," "cautious step," "quantum jump," "presidential backing,"* and *"Nixon's desire"*: Perry, 526–33. *"the influence he has had"*: Webster.

Falling in love: *Selling film:* Bonanos, 145–46; Estrin. **Quotations:** *"bottom line's in heaven"*: Wensberg, 229; McElheny, 420. *"principal cheerleader"*: Czach. *"He was boss"*: Blout, 47. *"hubris"*: Gonci.

Adams, Ansel. *Ansel Adams: An Autobiography.* Little, Brown, 1985.
Bernstein, Jeremy. "I Am a Camera." *NY Rev. Books* 35 (1988): 21.
Blout, Elkan. "Polaroid: Dreams to Reality." *Daedalus* 125 (1996): 39.
Bonanos, Christopher. *Instant: The Story of Polaroid.* Princeton, 2012.
Boyle, Willard, and George E. Smith. "Charge Coupled Semiconductor Devices." *Bell Sys. Tech. J.* 49 (1970): 587.
Burrows, William E. *Deep Black.* Berkley Books, 1986.
Business Week, "Love Is Ammunition for a Texas Airline." June 26, 1971.
Campbell, F. W. "Edwin Herbert Land." *Biog. Mem. Fell. Roy. Soc.* 40 (1994): 197.
Cordtz, Dan. "How Polaroid Bet Its Future on the SX-70." *Fortune*, Jan. 1974.
Czach, Elizabeth. "Polavision Instant Movies." *Moving Image* 2 (2002): 1.
Edison, Thomas. Patent Caveat 110: Peephole Kinetoscope; Motion picture cameras, filed Oct. 8, 1888.
Estrin, James. "Kodak's First Digital Moment." *NY Times*, Aug. 12, 2015.
Fantel, Hans. "Instant Movies: Shoot Now, See Now." *Pop. Mech.*, Aug. 1977.
Fierstein, Ronald K. *A Triumph of Genius.* Ankerwycke, 2015.
Fortune. "Dr. Land's Latest Fantasy." June 1972.
Gonci, Richard. "A Boston Story in 50 Words." *Boston Globe*, Mar. 17, 2017.
Harvard University. "Kennedy, Clarence. Papers and Photographs, 1921–1958."
Holt, D. D. "Three Living Leaders." *Fortune*, Mar. 23, 1981.
Kennedy, Clarence. "Photographing Art." *Magazine of Art*, Apr. 1937.
Land, Edwin H. "Experiments in Color Vision." *Sci. Am.* 200 (1959): 84.
———. "Chairman's Letter to Shareholders." *Polaroid Annual Report*, 1980.
Laurence, William L. "One-Step Camera Is Demonstrated." *NY Times*, Feb. 22, 1947.
Life. "Light Control: Polaroid Filters Make Enemy Targets Easier to See." Feb. 7, 1944.
———. "Unique Company Hits a Photographic Bonanza." Jan. 25, 1963.
Manchester, Harlan. *New World of Machines.* Random House, 1945.
———. "Pictures in 60 Seconds." *Sci. Am.*, April 1947.
McCune, William. Oral History, Concord Public Library, July 11, 1996.
McElheny, Victor K. *Insisting on the Impossible.* Perseus, 1998.
Middeldorf, Ulrich. "Clarence Kennedy 1892–1972." *Art Journal* 32 (1973): 372.
New York Times. "The Camera Does the Rest." Feb. 22, 1947.
———. "Dr. Kennedy Dies: Art Historian, 79." July 31, 1972.
Newsweek. "General Patton." July 26, 1943.
Norris, Pat. *Spies in the Sky.* Springer, 2008.

Ortner, Everett H. "Instant Movies." *Pop. Sci.*, July 1977.
Perry, Robert L. *A History of Satellite Reconnaissance*. Ed. James Outzen. NRO, 2012.
Richelson, Jeffrey. *The Wizards of Langley*. Westview, 2001.
Shumacher, Edward. "Polaroid Corp. Unveils Home Movie Camera." *Wash. Post*, Apr. 27, 1977.
Smith, F. Dow. "The Vision and Color World of Edwin Land." *Optics & Phot. News*, Oct. 1994, 30.
Smith, George E. Oral History, January 17, 2001.
TCP. "Report by the Technological Capabilities Panel." In *Foreign Relations of the United States, 1955–1957, Nat. Sec. Pol., Vol. XIX*. GPO, 1955.
Time. "Polaroid's Big Gamble on Small Cameras." June 26, 1972.
Tripsas, Mary, and Giovanni Gavetti. "Capabilities, Cognition, and Inertia: Evidence from Digital Imaging." *Strat. Man. J.* 21 (2000): 1147.
———."Photography: At Long Last, Land's Instant Movies." May 9, 1977.
Wang, Zuoyue. *In Sputnik's Shadow*. Rutgers, 2008.
Webster, William H. "Proposed Remarks by William H. Webster, Director of Central Intelligence, at the Security Affairs Support Association Dinner." Bolling Air Force Base, May 25, 1988.
Wensberg, Peter C. *Land's Polaroid*. Houghton Mifflin, 1987.

CHAPTER 5

Introduction and **Eight megabytes of sexual satisfaction:** *Factory:* Isaacson, 225; Schlender, 108–9; Stross, 124–25. *Businessland, NeXT sales:* Stross, 233–34, 274. *Sun:* Stross, 212–15, 258–59. *NeXT decline:* Linzmayer, 215; Pollack, "Quitting"; Stross, 218–20, 255, 274, 301–2, 329. **Quotations:** *"Jobs is back"* and *"Jobs is known for"*: Pollock, "Star." *"This is a revolution"* and *"One of my heroes"*: Jobs, NeXT Introduction. *"'wow' back"*: *Newsweek*. *"Vatican II"*: W. Smith. *"piss on it"*: Cringely, 311. *"write Sony a check"*: Hume. *"can of paint"*: Shore. *"sexual satisfaction"*: *San Francisco Examiner*. *"kick the shit"* and *"screaming"*: Stross, 210. *"risk our company"*: Jobs, NeXT Introduction. *"Biggest mistake I made"*: Stross, 301. *"Jobs is not one of them"*: Pitta, 137.

When Moses doubles down: *Exit from Apple:* Isaacson, 180–210; Schlender, 87–92; Sculley, 284–317. *Wozniak on early Apple: Wozniak*, 150–206. *Market share:* Reimer. *Bozos:* Sculley, 241. *Apple decline:* Schlender, 72–88; Sculley 227–91. *Fan:* Elliot, 30–31. *Get out of hardware:* Isaacson, 232; Stross, 310. **Quotations:** *"unbalancing the whole"*: Vail, 351. *"get his name"*: Sculley, 230.

Isaac Newton vs. Steve Jobs: *Newton and predecessors:* see chapter 5 note for *"launched Newton on the path"* on page 327. *Raskin:* Elliott; Hertzfeld; Isaacson, 108–13; Lammers, 227–45. **Quotations:** *"unsociable temper"*: Newton. *"shithead who sucks"*: Isaacson, 95; Linzmayer, 74. *"the story's too complicated"* and *"tablet"*: Schlender, 403–4.

Futureworld and **From Tubby to PIC:** *3D in Utah:* Catmull, 16–17; Price, 10–15; Rubin, 106–13. *Schure:* Price, 16–29; Rubin, 103–33. *Lucas:* Price, 30–35; Rubin, 137–41. *The PIC:* Catmull, 30; Linzmayer, 225–28. *Alan Kay:* Catmull, 39; Price, 64–66; Rubin, 298–99. *Jobs buys Pixar:* Catmull, 41–44; Linzmayer, 218–19; Price, 61, 72–74; Rubin, 411–13. **Quotations:** *"wasted two years"*: Rubin, 130. *"office of the future"*: Price, 20; A. Smith, 13–14. *"madman in Long Island"*: Rubin, 121. *"the house of Utah"*: A. Smith, 17. *"turns out to be a minuscule market"*: Miller. *"compete with Apple"*: Catmull, 41. *"after the divorce"*: Price, 67. *"How could GM"*: Price, 73. *"at the Movies"*: Wilson.

Fire-hydrant years and **Buzz and Woody save the day:** *Pixar stumbles, PIC falls:* Catmull, 53; Deutschman, 120–22; Linzmayer, 219–22; Price, 111–16. Toy Story, *Disney:* Catmull, x–xi, 55–56; Linzmayer, 220–23; Price, 69, 93–95, 117–39. *IPO:* Linzmayer, 222–23; Levy, 153–60. **Quotations:** *"will explode"*: Linzmayer, 219. *"PC industry in 1978"*: Wilson. *"ankle-deep"*: Kahney. *"more emotion and humor"*: Schlender, 169. *"visually astounding"*: Rechtshaffen. *"rebirth"*: Larsen. *"dawn"*: Ebert.

Movies and drugs and **Balancing Ugly Babies and the Beast:** *Genentech:* Hughes, 94–97; Robbins-Roth, 19–22. *Bond, monkey:* Broccoli, 126–78; Lycett, 393. **Quotations:** *"ill-shapen"*: Bacon, 387. *"Originality is fragile"*: Catmull, 131, 135. *"not even good enough"* and *"Limey truck driver"*: Broccoli, 128, 177.

How to win at chess and **Rescue operations:** *Candor at Pixar:* Catmull, 85–105. *Genentech publications:* Fraser. *Jobs 2.0:* Isaacson, 293–339; Schlender, 194–248. *iTunes:* Isaacson, 394–403. **Quotations:** *"I'm not a filmmaker"*: Schlender, 333. *"left-handed helix"*: Winslow. *"Attila the Hun"* and *"Los Alamos"*: Schlender, 222, 234. *"best innovation"*: Isaacson, 334. *"miracle"*: Baxter, 7.

The first three rules: *Xerox PARC:* Hiltzik; D. Smith. *Kodak:* Estrin. **Quotations:** *"innovation landfill"*: Elliot, 162. *"had to sandbag"*: Hiltzik, 264–65 (quoting John Ellenby). *"drought"*: Catmull, 130–31.

Bacon, Francis. *Francis Bacon: The Major Works*. Ed. Brian Vickers. Oxford, 2008.
Ball, W. W. Rouse. *A Short Account of the History of Mathematics*. Dover, 1960.
Baxter, James Phinney. *Scientists against Time*. Little, Brown, 1946.
Broccoli, Albert R., and Donald Zec. *When the Snow Melts*. Boxtree, 1998.
Catmull, Edwin E., and Amy Wallace. *Creativity, Inc*. Random House, 2014.
Cohen, I. Bernard. *The Newtonian Revolution*. Cambridge, 1980.
Cringely, Robert X. *Accidental Empires*. Addison-Wesley, 1992.
Deutschman, Alan. *The Second Coming of Steve Jobs*. Broadway, 2000.
Ebert, Roger. "Toy Story." *RoberEbert.com*, Nov. 22, 1995.
Elliot, Jay, and William Simon. *The Steve Jobs Way*. Vanguard, 2011.
Elliott, Andrea. "Jef Raskin, 61, Developer of Apple Macintosh, Is Dead." *NY Times*, Feb. 28, 2005.
Estrin, James. "Kodak's First Digital Moment." *NY Times*, Aug. 12, 2015.

Fraser, Laura. "The Paper." Genentech: web.archive.org/web/*/www.gene.com/stories/the-paper.

Gal, Ofer, and Raz Chen-Morris. *Baroque Science*. U. Chicago, 2013.

Hall, A. Rupert. *Isaac Newton: Adventurer in Thought*. Cambridge, 1992.

Hertzfeld, Andy. *Revolution in the Valley*. O'Reilly, 2004.

Hiltzik, Michael. *Dealers of Lightning*. Harper, 1999.

Hooke, Robert. *Philosophical Experiments and Observations of the Late Eminent Dr. Robert Hooke*. W. Derham, 1726.

Hughes, Sally Smith. *Genentech*. U. Chicago, 2011.

Hume, Brit. "Steve Jobs Pulls Ahead of Microsoft Rival in Race for PC Supremacy." *Wash. Post*, Oct. 31, 1988.

Inwood, Stephen. *The Man Who Knew Too Much*. Macmillan, 2002.

Isaacson, Walter. *Steve Jobs*. Simon and Schuster, 2011.

Jardine, Lisa. *The Curious Life of Robert Hooke*. Perennial, 2005.

Jobs, Steve. Video: The NeXT Introduction. San Francisco, Oct. 12, 1988.

Kahney, Leander. "The Wilderness Years." *Newsweek* Oct. 10, 2011, 20.

Kasparov, Garry, and Mig Greengard. *How Life Imitates Chess*. Bloomsbury, 2007.

Lammers, Susan M. *Programmers at Work*. Tempus, 1986.

Larsen, Josh. "Toy Story." *Larsen on Film*, June 10, 2010.

Levy, Lawrence. *To Pixar and Beyond*. Houghton Mifflin Harcourt, 2016.

Linzmayer, Owen W. *Apple Confidential 2.0*. No Starch Press, 2004.

Lycett, Andrew. *Ian Fleming*. 1995.

Miller, Michael W. "Producers of Computer Graphics for Hollywood Find New Opportunities in Science and Industry." *WSJ*, Sep. 16, 1985.

Nauenberg, Michael. "Robert Hooke's Seminal Contribution to Orbital Dynamics." *Phys. Persp.* 7 (2005): 4.

Newsweek. "Mr. Chips: Steve Jobs Puts the 'Wow' Back in Computers." Oct. 24, 1988.

Newton, Isaac. Letter to Edmond Halley, June 20, 1686.

Pitta, Julia. "The Steven Jobs Reality Distortion Field." *Forbes*, Apr. 29, 1991, 137.

Pollack, Andrew. "The Return of a Computer Star." *NY Times*, Oct. 13, 1988.

———. "A Co-Founder of Next Is Quitting the Company." *NY Times*, May 4, 1991.

Price, David A. *The Pixar Touch*. Knopf, 2008.

Rechtshaffen, Michael. "Toy Story." *Hollywood Reporter*, Nov. 20, 1995.

Reimer, Jeremy. "Total Share: 30 Years of Personal Computer Market Share Figures." *Ars Technica*, Dec. 15, 2005.

Robbins-Roth, Cynthia. *From Alchemy to IPO*. Perseus, 2000.

Rubin, Michael. *Droidmaker*. Triad, 2005.

San Francisco Examiner. "Eight Megabytes of Sexual Satisfaction." Oct. 16, 1988.

Schlender, Brent, and Rick Tetzeli. *Becoming Steve Jobs*. Crown, 2015.

Sculley, John, and John A. Byrne. *Odyssey*. Fontana, 1987.

Shore, Joel, and Kristen Hedlund. "NeXT Pulls No Punches." *Comp. Resell. News*, Dec. 4, 1989.

Smith, Alvy Ray. "Digital Paint Systems." *IEEE Ann. Hist. Comp.* 23 (2001): 4.
Smith, Douglas K., and Robert C. Alexander. *Fumbling the Future.* William Morrow, 1988.
Smith, Wes. "The Cult of Steve." *Chicago Tribune*, Oct. 23, 1988.
Stross, Randall E. *Steve Jobs and the NeXT Big Thing.* Macmillan, 1993.
Vail, Theodore Newton. *Views on Public Questions: A Collection of Papers and Addresses of Theodore Newton Vail, 1907–1917.* Priv. print., 1917.
Whiteside, D. T. "Before the Principia." *J. Hist. Astr.* 1 (1970): 5.
Wilson, John W. "Look What Steve Jobs Found at the Movies." *Bus. Week*, Feb. 17, 1986, 37.
Winslow, Ron. "Genentech's Levinson Sets the Record Straight on DNA." *WSJ*, Jan. 14, 2009.
Wozniak, Steve, and Gina Smith. *iWoz.* Norton, 2006.

INTERLUDE

Death of Smith: Rae, 434–35; Ross, 434–36. *Misinterpretations:* see Interlude note for *"prized his works on ethics"* on page 328. *Smith and Newton:* Hetherington. *The invisible hand:* Kennedy; Rothschild, 116–56; Wight. *Phase transitions, emergence overviews:* Ball; Gell-Mann; Laughlin; Solé; Strogatz. **Quotations:** *"notably rare exceptions"*: Greenspan. *"Germany largely at peace"*: Buchanan, 45. *"ultimate laws"*: Greene, 373. *"cliché not discussed"*: Laughlin, "Theory," 30. *"go to another world"*: Rae, 435. *"invisible chains"* and *"greatest discovery"*: Smith, "Astronomy," 45, 105. *"mildly ironic joke"*: Kennedy, 239; Rothschild, 116.

CHAPTER 6

Introduction; Jane Austen, physicist; and **Phase diagrams:** *Traffic, popular:* Ball, 156–77. *Traffic, technical reviews:* Helbing; Nagel. *Nagoya Dome:* Tadaki. *Traffic interventions:* Treiber, 403–22. *Turbulence, golf:* Tsinober; Grinham. **Quotation:** *"understand simple things"*: Feynman, 230.

From gas masks to forest fires, How to be simple, and **Six degrees of Kevin Cricket:** *Percolation, history:* Bacaër, 121–26; Broadbent; Grimmett; Hammersley; Hammersley and Morton (Druid circles); Kendall. *Percolation, epidemics:* reviewed in Newman, 591–675; Pastor-Satorras. *Forest fires:* Hantson (models); Malamud (Yellowstone); Scott, 349 (slope); Sullivan (models); Weir (humidity); Zinck (models). *Small world:* Strogatz, 233–48; Watts, *Degrees*, 31–42, 69–74, 93–95. *Networks, reviewed:* Newman. **Quotation:** *"The old-timer"*: Scott, 342.

The power in the tail and **When terror goes viral:** Interview with Neil Johnson. *Fat tails, reviews:* Gabaix; Farmer; Johnson, *Financial*; Mandelbrot; Sornette. *Conflict* *data:* Bohorquez (insurgent conflicts); Clauset (terror events); Johnson, "Online Ecology"; Radicchi.

Anderson, Philip W. *More and Different.* World Scientific, 2011.
Bacaër, Nicolas. *A Short History of Mathematical Population Dynamics.* Springer, 2011.
Ball, Philip. *Critical Mass.* Farrar, Straus and Giroux, 2004.
Bohorquez, Juan C., et al. "Common Ecology Quantifies Human Insurgency." *Nature* 462 (2009): 911.
Broadbent, S. R., and J. M. Hammersley. "Percolation Processes." *Math. Proc. Camb. Phil. Soc.* 53 (1957): 629.
Brown, Laurie M., and Tian Yu Cao. "Spontaneous Breakdown of Symmetry." *Hist. Stu. Phys. Bio. Sci.* 21 (1991): 35.
Brown, Timothy J., et al. *Coarse Assessment of Federal Wildland Fire Occurrence Data.* Nat. Wildfire Coordinating Group, 2002.
Buchanan, Mark. *Forecast.* Bloomsbury, 2013.
Carroll, Sean M. *The Particle at the End of the Universe.* Dutton, 2012.
Clauset, Aaron, et al. "On the Frequency of Severe Terrorist Events." *J. Conflict Resolution* 51 (2007): 58.
Cooper, Leon N., and Dimitri Feldman, eds. *BCS: 50 Years.* World Scientific, 2011.
Farmer, J. Doyne, and John Geanakoplos. "The Virtues and Vices of Equilibrium and the Future of Financial Economics." *Complexity* 14 (2009): 11.
Feynman, Richard P. *Perfectly Reasonable Deviations.* Basic Books, 2005.
Gabaix, Xavier. "Power Laws in Economics: An Introduction." *J. Econ. Persp.* 30 (2016): 185.
Gell-Mann, Murray. *The Quark and the Jaguar.* Macmillan, 1994.
Greene, Brian. *The Elegant Universe.* Norton, 1999.
Greenspan, Alan. "How Dodd-Frank Fails to Meet the Test of Our Times." *Financial Times*, Mar. 30, 2011.
Grimmett, Geoffrey, and Dominic Welsh. "John Michael Hammersley." *Biog. Mem. Fell. Roy. Soc.* 53 (2007): 163.
Grinham, T. "How Do Dimples on Golf Balls Affect Their Flight?" *Sci. Am.* 290 (2004): 111.
Hammersley, J. M. "Origins of Percolation Theory." In G. Deutscher et al., *Percolation Structures and Processes*, Ann. Israel Phys. Soc. 5 (1983), 47.
Hammersley, J. M., and K. W. Morton. "Poor Man's Monte Carlo." *J. Royal Stat. Soc. B* 16 (1954): 23.
Hantson, Stijn, et al. "Global Fire Size Distribution." *Int. J. Wildland Fire* 25 (2016): 403.
Helbing, Dirk. "Traffic and Related Self-Driven Many-Particle Systems." *Rev. Mod. Phys.* 73 (2001): 1067.
Hetherington, Norriss S. "Isaac Newton's Influence on Adam Smith's Natural Laws in Economics." *J. Hist. Ideas* 44 (1983): 497.
Hoddeson, Lillian, ed. *The Rise of the Standard Model.* Cambridge, 1997.

Johnson, Neil F., et al. "New Online Ecology of Adversarial Aggregates: ISIS and Beyond." *Science* 352 (2016): 1459.
———. *Financial Market Complexity*. Oxford, 2003.
Kendall, David. "Toast to John Hammersley." In *Disorder in Physical Systems*, edited by G. Grimmett and D. Welsh., 1. Oxford, 1990.
Kennedy, Gavin. "Adam Smith and the Invisible Hand." *Econ. Journal Watch* (2009): 63.
Krugman, Paul. *The Self-Organizing Economy*. Blackwell, 1996.
Laughlin, R. B., and David Pines. "The Theory of Everything." *PNAS* 97 (2000): 28.
———. *A Different Universe*. Basic Books, 2005.
Lederman, Leon, and Dick Teresi. *The God Particle*. Houghton Mifflin Harcourt, 1993.
Malamud, Bruce D., et al. "Forest Fires: An Example of Self-Organized Critical Behavior." *Science* 281 (1998): 1840.
Mandelbrot, Benoit B., and Richard L. Hudson. *The (Mis)Behavior of Markets*. Basic Books, 2004.
McLean, Iain. *Adam Smith: Radical and Egalitarian*. Edinburgh, 2006.
Mirowski, Philip. *More Heat Than Light*. Cambridge, 1989.
Montes, Leonidas. "Newtonianism and Adam Smith." In *The Oxford Handbook of Adam Smith*, edited by C. Berry et al., 36. Oxford, 2013.
Nagel, Kai, et al. "Still Flowing: Approaches to Traffic Flow and Traffic Jam Modeling." *Op. Res.* 51 (2003): 681.
Newman, M. E. J. *Networks: An Introduction*. Oxford, 2010.
Pastor-Satorras, R., et al. "Epidemic Processes in Complex Networks." *Rev. Mod. Phys.* 87 (2015): 925.
Radicchi, Filippo, and Claudio Castellano. "Leveraging Percolation Theory to Single out Influential Spreaders in Networks." *Phys. Rev. E* 93 (2016).
Rae, John. *Life of Adam Smith*. Macmillan, 1895.
Romilly, Sir Samuel. "Letter LXXI, To Madam G—," Aug. 20, 1790, in *Memoirs of the Life of Sir Samuel Romilly*, edited by His Sons, Vol. I (1840), 404.
Ross, Ian S. *The Life of Adam Smith*. Oxford, 2010.
Rothschild, Emma. *Economic Sentiments*. Harvard, 2001.
Schofield, Robert E. "An Evolutionary Taxonomy of Eighteenth-Century Newtonianisms." *Stu. Eighteenth-Cent. Cult.* 7 (1978): 175.
Scott, Andrew C., et al. *Fire on Earth: An Introduction*. Wiley, 2014.
Smith, Adam. *An Inquiry into the Nature and Causes of the Wealth of Nations*. Glasgow Ed., Vol. 2. Oxford, 1976.
———. "History of Astronomy." In *Essays on Philosophical Subjects*. Glasgow Ed., Vol. 3. Oxford, 1980.
Solé, Ricard V. *Phase Transitions*. Princeton, 2011.
Sornette, Didier. "Physics and Financial Economics (1776–2014)." *Rep. Prog. Phys.* 77 (2014).
Strogatz, Steven H. *Sync: The Emerging Science of Spontaneous Order*. Hyperion, 2003.

Sullivan, A. L. "A Review of Wildland Fire Spread Modelling, 1990–2007. 2: Empirical and Quasi-Empirical Models." *Int. J. Wildland Fire* 18 (2009): 369.
Tadaki, Shin-ichi, et al. "Phase Transition in Traffic Jam Experiment on a Circuit." *New J. of Physics* 15 (2013).
Treiber, Martin, and Arne Kesting. *Traffic Flow Dynamics*. Springer, 2013.
Tsinober, A. *An Informal Conceptual Introduction to Turbulence*. Springer, 2009.
Watts, Duncan J. *Six Degrees: The Science of a Connected Age*. Norton, 2003.
Watts, Duncan J., and Steven H. Strogatz. "Collective Dynamics of 'Small-World' Networks." *Nature* 393 (1998): 440.
Weir, John. "Probability of Spot Fires during Prescribed Burns." *Fire Mgmt. Today* 64, no. 2 (2004): 24.
Wight, Jonathan B. "The Treatment of Smith's Invisible Hand." *J. Econ. Educ.* 38 (2007): 341.
Witten, Edward. "Phil Anderson and Gauge Symmetry Breaking." In *PWA90*, edited by P. Chandra et al., WSPC, 2016.
Zinck, Richard D., and Volker Grimm. "Unifying Wildfire Models from Ecology and Statistical Physics." *Am. Naturalist* 174 (2009): E170.

CHAPTER 7

Mormons, murder, and monkeys: *Assassination:* Roberts, *VII*, 99–109; Wicks. *Mormonites:* Bowman, xiv. *Missouri militia:* Roberts, *III*, 202–4. *Conspiracy, trial:* Wicks, 157–80, 216–21, 233. *Young's revelation:* Young, 170. *To Utah, 150 at a time:* Data on company sizes from Church Historian's Press. *Monkey brains and magic numbers:* Bennett (Morin); Fost, 64 (Minerva); Dunbar. *Criticism:* Andrew. **Quotations:** *"resigned to my lot"*: Roberts, *VI*, 605. *"to revive us"* and *"our feelings at the time"*: Roberts, *VII*, 101. *"Cast behind you"*: Brodhead, 57 (citing Emerson 1838 address). *"exterminated"*: Roberts, *III*, 192. *"the right place"*: Ostling, 44. *"To be groomed"*: Dunbar, *Grooming*, 1. *"smallest independent units"*: Dunbar, "Constraint," 686. *"parking spaces"*: Gladwell, 185.

The magic number: *Management trends:* O'Leonard; Rajan.

Andrew, R. J., et al. "Open Peer Commentary on 'Coevolution of Neocortical Size, Group Size and Language in Humans.'" *Behav. Brain Sci.* 16 (1993): 681.
Bennett, Drake. "The Dunbar Number." *Bloomberg BusinessWeek*, Jan. 14, 2013: 52.
Bowman, Matthew Burton. *The Mormon People*. Random House, 2012.
Brodhead, Richard H. "Prophets in America ca. 1830." *J. Mormon Hist.* 29 (2003): 43.
Church Historian's Press. "Brigham Young Vanguard Company (1847)." *Mormon Pioneer Overland Travel, 1847–1868*.
De Vany, Arthur S., and W. David Walls. "Motion Picture Profit." *J. Econ. Dyn. Control* 28 (2004): 1035.
Dunbar, Robin. "Neocortex Size as a Constraint on Group Size in Primates." *J. Hum. Evol.* 22 (1992): 469.

———. "Coevolution of Neocortical Size, Group Size and Language in Humans." *Behav. Brain Sci.* 16 (1993): 681.
———. *Grooming, Gossip, and the Evolution of Language.* Harvard, 1996.
Epstein, Edward Jay. *The Hollywood Economist 2.0.* Melville, 2012.
Fost, Joshua. "New Look at General Education." In *Building the Intentional University*, edited by Stephen M. Kosslyn and Ben Nelson. MIT, 2017.
Gladwell, Malcolm. *The Tipping Point.* Little, Brown, 2000.
Leipzig, Adam. "Sundance 2014." *Cultural Weekly*, Jan. 22, 2014.
O'Leonard, Karen, and Jennifer Krider. "Leadership Development Factbook 2014." Bersin by Deloitte, 2014.
Ostling, Richard N., and Joan K. Ostling. *Mormon America.* HarperCollins, 2007.
Rajan, Raghuram G., and Julie Wulf. "The Flattening Firm." *Rev. Econ. Stat.* 88 (2006): 759.
Roberts, Brigham H. *History of the Church of Jesus Christ of Latter-day Saints.* Deseret News, Vol. III (1905); Vol. VI (1912); Vol. VII (1932).
Sparviero, Sergio. "Hollywood Creative Accounting." *Media Ind. J.* 2 (2015).
Wicks, Robert Sigfrid, and Fred R. Foister. *Junius and Joseph.* Utah State, 2005.
Wong, Chi H., et al. "Estimation of Clinical Trial Success Rates and Related Parameters." *Biostat* (2018).
Young, Brigham. *The Complete Discourses of Brigham Young.* Ed. R. S. Van Wagoner. Vol. 1. Smith-Pettit, 2009.

CHAPTER 8

Web of DARPA: *NSF, NIH:* see chapter 1 note for *"epoch-making"* on page 319. *Sputnik:* Brzezinski, 145–47, 176; Drury (Teller); *Newsday*; Schwartz. *McElroy, DARPA:* Daye; Hafner, 13–24. The name of the organization changed between ARPA and DARPA three times (1972, 1993, 1996). For simplicity I use the current name in all instances. **Quotations:** *"go out of circulation"*: Bush, 63. *"freeway overpasses"*: Weinberger, 34. *"our survival"*: Mieczkowski, 16. *"gravest danger"*: Roberts. *"lots of soap"* and *"the proposals are suggestions"*: Hafner, 14, 19.

A giant nuclear suppository: *Christofilos:* Jacobsen, 66–69; Weinberger, *Imagineers*, 94–97. *Mechanical elephants and more:* Mervis; Meyer; Taleyarkhan; Weinberger, *Imagineers*, 204; Weinberger, "Scary Things." *Created new disciplines:* DARPA, "Breakthrough Technologies"; Weinberger, *Imagineers*, 98–103 (seismology). *DARPA, Utah, Xerox PARC, Engelbart:* Catmull, 11–13; Rubin, 106–7; Weinberger, *Imagineers*, 121; Hern; Hiltzik, 14–18; Smith, 61–78, 87–88. **Quotations:** *"suppository"* and *"bunch of incompetents"*: Weinberger, *Imagineers*, 96, 101. *"profoundly influenced"*: Catmull, 13. *"developed at DARPA"*: Hiltzik, 145.

Six degrees of red balloons and **A toothpaste problem:** *ARPANET:* Cerf. *Balloon Challenge:* Pickard; Tang; Trewhitt; interview with Doug Wickert. *DARPA structure:* Dugan; Mervis; Travis. *Coors, Kraft:* Martinez, 63–73, 139–53. **Quotations:** *"fueled by* *donuts"*: Trewhitt. *"a bit of colored ribbon"*: attributed to Napoleon Bonaparte, speaking to the captain of HMS *Bellerophon* on July 15, 1815.

A shredding problem: *Dead Sea Scrolls:* Gell-Mann, 323. *Academic studies:* Bloom; Ordóñez; Wade. *Coughran:* Hill; Bock, 388 (180 reports); interview with Bill Coughran. *Taylor, PARC:* Smith, 76–79; Hiltzik, 150–53. **Quotations:** *"increased [wage] dispersion"*: Wade, 528. *"rather be worth 100 million"*: Bandiera, 625. *"keeping the reins"*: Hill. *"wasn't the best decision"*: interview. *"tended to wither"*: Hiltzik, 152. *"organizational distractions"*: Smith, 77 (both citing Charles Thacker, leader of the Alto personal computer project; coinventor of ethernet).

Postscript: See chapter 8 notes for *"jail terms"* on page 333 and *"for both types of deliveries"* on page 333.

Allin, Sara, et al. "Physician Incentives and the Rise in C-Sections." *NBER Working Paper*, Mar. 1, 2015.
Ariely, Dan. *Predictably Irrational.* Rev. ed. HarperCollins, 2009.
Bandiera, Oriana, et al. "Matching Firms, Managers, and Incentives." *J. Labor Econ.* 33 (2015): 623.
Bersin & Associates. "High-Impact Leadership Development," 2011.
Bloom, Matt, and John G. Michel. "The Relationships among Organizational Context, Pay Dispersion, and Managerial Turnover." *Acad. Mgmt.* 45 (2002): 33.
Bock, Laszlo. *Work Rules!* Twelve, 2015.
Brzezinski, Matthew. *Red Moon Rising.* Times Books, 2007.
Bush, Vannevar. *Pieces of the Action.* Morrow, 1970.
Catmull, Edwin E., and Amy Wallace. *Creativity, Inc.* Random House, 2014.
Cerf, Vinton G. "The Day the Internet Age Began." *Nature* 461 (2009): 1202.
Csaszar, Felipe A. "An Efficient Frontier in Organization Design." *Org. Sci.* 24 (2013): 1083.
DARPA. "Breakthrough Technologies for National Security," Mar. 2015.
Daye, Derrick. "Neil McElroy Memo." *Branding Strategy Insider*, June 12, 2009.
Drury, Allen. "Missiles Inquiry Will Open Today." *NY Times*, Nov. 25, 1957.
Dugan, Regina E., and Kaigham J. Gabriel. "'Special Forces' Innovation." *Harv. Bus. Rev.*, Oct. 1, 2013.
Gell-Mann, Murray. *The Quark and the Jaguar.* Macmillan, 1994.
Hafner, Katie, and Matthew Lyon. *Where Wizards Stay Up Late.* 1998. Touchstone, 1996.
Hern, Daniela. "The Mother of All Demos, 1968." *WIRED*, Dec. 13, 2013.
Hill, Linda A., et al. "Collective Genius." *Harv. Bus. Rev.*, June 2014.
Hiltzik, Michael A. *Dealers of Lightning.* HarperCollins, 1999.
Jacobsen, Annie. *The Pentagon's Brain.* Little, Brown, 2015.
Kahneman, Daniel. *Thinking, Fast and Slow.* Farrar, Straus and Giroux, 2013.
Levitt, Steven D., and Stephen J. Dubner. *Freakonomics.* Rev. ed. William Morrow, 2005.
Martinez, Marian Garcia, ed. *Open Innovation in the Food and Beverage Industry.* Woodhead, 2013.
Mervis, Jeffrey. "What Makes DARPA Tick?" *Science* 351 (2016): 549.
Meyer, Josh. "Trading on the Future of Terror." *LA Times*, July 29, 2003.

Mieczkowski, Yanek. *Eisenhower's Sputnik Moment*. Cornell, 2013.
Newsday. "Russia Wins Space Race." Oct. 5, 1957.
NPW (National Partnership for Women and Families). "Why Is the US Cesarean Section Rate So High?" Aug. 2016.
Ordóñez, Lisa, et al. "Goals Gone Wild." *Acad. Mgmt. Persp.* 23 (2009): 6.
Pickard, Galen, et al. "Time-Critical Social Mobilization." *Science* 334 (2011): 509.
Roberts, Chalmers M. "Enormous Arms Outlay Is Held Vital to Survival." *Wash. Post and Times Herald*, Dec. 20, 1957.
Rubin, Michael. *Droidmaker*. Triad, 2006.
Sah, Raaj K., and Joseph E. Stiglitz. "The Architecture of Economic Systems: Hierarchies and Polyarchies." *Amer. Econ. Rev.* 76 (1986): 716.
Sakala, Carol, et al. "Maternity Care and Liability." *Women's Health Issues* 23 (2013): e7.
Schwartz, Harry. "A Propaganda Triumph." *NY Times*, Oct. 6, 1957.
Smith, Douglas K., and Robert C. Alexander. *Fumbling the Future*. William Morrow, 1988.
Taleyarkhan, R. P., et al. "Evidence for Nuclear Emissions during Acoustic Cavitation." *Science* 295 (2002): 1868.
Tang, John, et al. "Reflecting on the DARPA Red Balloon Challenge." *Comm. ACM* 54 (2011): 78.
Thaler, Richard H. *Misbehaving: The Making of Behavioral Economics*. Norton, 2016.
Travis, John. "Interview with Michael Goldblatt." *Biosec. and Bioterr.* 1 (2003): 155.
Trewhitt, Ethan. Accessed July 20, 2018. https://cacm.acm.org/blogs/blog-cacm/76324-preparing-for-the-darpa-network-challenge.
Wade, James, et al. "Overpaid CEOs and Underpaid Managers: Fairness and Executive Compensation." *Org. Sci.* 17 (2006): 527.
Weinberger, Sharon. "Scary Things Come in Small Packages." *Wash. Post*, Mar. 28, 2004.
——. *The Imagineers of War*. Knopf, 2017.

CHAPTER 9

The Needham Question: For a popular overview of Needham's decades-long relationships with Lu and with China, see Winchester, the source for the opening anecdote. For a bibliography on Needham and the Needham Question, see Nathan Sivin's entry in the Oxford Bibliographies. *Lu and Needham:* Winchester, 34–57 (citing Needham's diaries); Lu, 2–8, 24–25, 29–34. *Needham Question, reviewed:* Finlay; Sivin, "Revolution." *GDP:* Maddison, 379–81. *Civil service:* Elman. *Literacy:* R. Allen, 25; Mokyr, 292. *Navy:* Dreyer; Hobson, 140–48; Morris, 16, 517. *India:* Metcalf, 29–91; Walsh, 100–136. **Quotations:** *"When you have learned"*: Feynman, 230. *"bushy white beard"*: Lu, 2. *"almost a lisp"* and *"why not develop"*: Winchester, 37, 57. *"greatest single act"*: Finlay, 265. *"so much like my own"*: Needham, "Foreword," xi. *"There is nothing we lack"*: Maddison, 164.

Eight minutes that changed the world and **Three conditions for a loonshot nursery:** *Earth's motion:* Dutta; Eastwood; Linton, 24–39, 115–22; Padmanabhan, 5–13; Ragep; Ramasubramanian; Weinberg, 66–72, 77–86, 132–40. *Copernicus, Church:* Koestler,

144–53; Westman, 133–34, 197. *Kepler:* Gingerich, "Kepler"; Koestler; Voelkel. *Hooke:* see chapter 5 note for "*launched Newton on the path*" on page 327. *Growth data:* Maddison, 72, 376, 382. *Islamic science:* Al-Khalili; Lindberg, 27–167; Ragep; Saliba; see chapter 9 note for "*no Copernican theory*" on page 336. **Quotations:** "*as being absurd*": Voelkel, 22. "*a little lap-dog*": Koestler, 236. "*For the first time*": Kepler, "Letter" (trans: Baumgardt, 31–32). "*exists only in the mind*": Kepler, *Astronomy*, 234 (trans. Koestler, 319). "*These eight minutes*": Kepler, *Astronomy*, 286. "*kindred spirit*": Cohen, "Kepler," 27. "*freeing himself*": Einstein, 226.

Movies, **Drugs**, and **The fate of empires:** *Rise of Hollywood:* Easton, 21–23; Gabler; Hampton 7–13, 71–79; Gil; Russell, 237–39. *Leonard, insulin:* Banting; Bliss. *Drug discovery, biotech histories:* Hughes; Robbins-Roth; Sneader. *Tycho:* Christianson; Thoren. *Shen:* Sivin, "Shen"; Sun, 21–80; Zuo; Needham, vol. 3, 135–45, 262, 415–35, 603–18. *Shen project and assistant:* Sun, 61–69. **Quotations:** "*No other invention*": Hampton, 13. "*useless and harmful*": F. Allen, 813–15. "*he felt stronger*": Banting, 144. "*It was a resurrection*": Kienast, 14–15. "*the most interesting character*": Needham, vol. 1, 135. "*On certain wooden shelves*": Needham, vol. 3, 482. "*audacity*" and "*our equal*": Thoren, 380. "*Since I retired*": Sivin, "Shen," 10; Zuo, 211 (Sivin translates the final phrase as "to chat with" rather than "to talk to").

Loonshot life support and **Why England:** *Wittenberg school:* Voelkel, 22; Westman, 141–70. *Kerala school:* Joseph, 372–444; Plofker, 217–253. *Royal Society:* Gribbin; Sprat. *Boyle, Papin, pump:* H. F. Cohen, 111–25; Papin; Shapin, *Air-Pump*, 274–76; Shapin, *Truth*, 356–58; Wootton, 491–93, 499–508. **Quotations:** "*Science was to be fostered*": Merton, 234. "*It is doubtful*": Cohen, *Modern Science*, 72. "*extraordinary Inventions . . . Glory of the Western World*": Sprat, 74–79.

Acemoglu, Daron, and James Robinson. *Why Nations Fail*. Crown, 2012.

Alito, Samuel A., Jr. "The Origin of the Baseball Antitrust Exemption." *J. Supreme Court Hist.* 34 (2009): 183.

Al-Khalili, Jim. *The House of Wisdom*. Penguin, 2011.

Allen, Frederick M. *Studies Concerning Glycosuria and Diabetes*. Harvard, 1913.

Allen, Robert C. *Global Economic History*. Oxford, 2011.

Banting, F. G., et al. "Pancreatic Extracts in the Treatment of Diabetes Mellitus." *Can. Med. Assoc.* 12 (1922): 141.

Baumgardt, Carola. *Johannes Kepler: Life and Letters*. Philosophical Library, 1951.

Bliss, Michael. *The Discovery of Insulin*, 25th Anniv. Ed. U. Chicago, 2007.

Brandt, Loren, et al. "From Divergence to Convergence: Reevaluating the History behind China's Economic Boom." *J. Econ. Lit.* 52 (2014): 45.

Christianson, J. R. *On Tycho's Island*. Cambridge, 2000.

Cohen, H. Floris. "The Rise of Modern Science as a Fundamental Pre-Condition for the Industrial Revolution." *Öst. Zeit. Ges.* 20 (2009): 107.

Cohen, I. Bernard. "Kepler's Century: Prelude to Newton's." *Vistas in Astr.* 18 (1975): 3.

———. "Introduction." In *Puritanism and the Rise of Modern Science*, edited by I. B. Cohen, 1–111. Rutgers, 1990.

Daly, Jonathan W. *Historians Debate the Rise of the West*. Routledge, 2014.

———. *The Rise of Western Power*. Bloomsbury, 2014.

Diamond, Jared M., and James A. Robinson, eds. *Natural Experiments of History*. Harvard, 2011.

Dreyer, Edward L. *Zheng He*. Pearson Longman, 2007.

Dutta, Amartya. "Āryabhata and Axial Rotation of Earth." *Resonance* 11 (2006): 51.

Easton, Carol. *The Search for Sam Goldwyn*. U. Mississippi, 2014.

Eastwood, Bruce, and Hubert Martin. "Michael Italicus and Heliocentrism." *Greek, Roman, Byz. Stud.* 27 (1986): 223.

Einstein, Albert. *Out of My Later Years*. Philosophical Library, 1950.

Elman, Benjamin. *A Cultural History of Civil Examinations in Late Imperial China*. U. California, 2000.

Feynman, Richard P. *Perfectly Reasonable Deviations*. Basic Books, 2005.

Finlay, Robert. "China, the West, and World History." *J. World Hist.* 11 (2000): 265.

Gabler, Neal. *An Empire of Their Own*. Crown, 1988.

Gil, Alexandra. "Breaking the Studios." *NYU J. Law & Liberty* 3 (2008): 83.

Gingerich, Owen. "The Great Martian Catastrophe and How Kepler Fixed It." *Phys. Tod.* 64 (2011): 50.

Gingerich, Owen, and Robert S. Westman. "The Wittich Connection." *Trans. Am. Phil. Soc.* 78 (1988): i.

Goldstone, Jack A. *Why Europe?* McGraw-Hill, 2009.

Golinski, Jan. *British Weather and the Climate of Enlightenment*. U. Chicago, 2007.

Gribbin, John. *The Fellowship*. Allen Lane, 2005.

Hampton, Benjamin Bowles. *A History of the Movies*. Covici, Friede, 1931.

Hobson, John M. *The Eastern Origins of Western Civilization*. Cambridge, 2004.

Hodgson, Marshall G. S. *The Venture of Islam*. U. Chicago, 1974.

Hughes, Sally Smith. *Genentech*. U. Chicago, 2011.

Jacob, Margaret C. *Scientific Culture and the Making of the Industrial West*. Oxford, 1997.

Jaramillo, Laura, and Cemile Sancak. "Why Has the Grass Been Greener on One Side of Hispaniola?" *IMF Staff Papers* 56 (2009): 323.

Joseph, George G. *The Crest of the Peacock*. 3rd ed. Princeton, 2011.

Kepler, Johannes. *New Astronomy*. Trans. William H. Donahue. Cambridge, 1992.

———. "To the Baron von Herberstein and the Estates of Styria," May 15, 1596.

Kienast, Margate. "I Saw a Resurrection." *Sat. Eve. Post* 211 (July 2, 1938): 14.

Koestler, Arthur. *The Sleepwalkers*. Macmillan, 1959.

Lin, Justin Y. *Demystifying the Chinese Economy*. Cambridge, 2012.

Lindberg, David C., and Michael H. Shank. *The Cambridge History of Science: Vol. 2, Medieval Science*. Cambridge, 2013.

Linton, C. M. *From Eudoxus to Einstein*. Cambridge, 2004.

Lu, Gwei-Djen. "The First Half-Life of Joseph Needham." In *Explorations in the History of Science and Technology in China*, edited by G. Li et al. Shanghai, 1982, 1–38.

Lunde, Paul, and Zayn Bilkadi. "Arabs and Astronomy." *Saudi Aramco World*, Jan./Feb. 1986, 4.
Maddison, Angus. *Contours of the World Economy, 1–2030 AD.* Oxford, 2007.
McClintick, David, and Anne Faircloth. "The Predator." *Fortune*, July 9, 1996.
Merton, Robert K. *The Sociology of Science.* U. Chicago, 1973.
Metcalf, Barbara D., and Thomas R. Metcalf. *A Concise History of Modern India.* Cambridge, 2006.
Mokyr, Joel. *A Culture of Growth: The Origins of the Modern Economy.* Princeton, 2016.
Morris, Ian. *Why the West Rules—For Now.* Profile, 2010.
Needham, Joseph. *Science and Civilization in China.* Cambridge, 1954–2015.
———. "Foreword." In Edgar Zilsel, *The Social Origins of Modern Science.* Kluwer, 2003.
Padmanabhan, T., et al., eds. *Astronomy in India.* Springer, 2010.
Papin, Denis. *A Continuation of the New Digester of Bones: Its Improvements, and New Uses It Hath Been Applyed to, Both for Sea and Land: Together with Some Improvements and New Uses of the Air-Pump.* J. Streater, 1687.
Plofker, Kim. *Mathematics in India.* Princeton, 2009.
Ragep, F. Jamil. "Tūsī and Copernicus: The Earth's Motion in Context." *Sci. Context* 14 (2001): 145.
———. "Copernicus and His Islamic Predecessors." *Hist. Sci.* 45 (2007): 65.
Ramasubramanian, K., et al. "Modification of the Earlier Indian Planetary Theory by the Kerala Astronomers." *Curr. Sci.* 66 (1994): 784.
Robbins-Roth, Cynthia. *From Alchemy to IPO.* Perseus, 2000.
Roston, Tom. "'Slumdog Millionaire' Shoot Was Rags to Riches." *Hollyw. Rep.*, Nov. 4, 2008.
Russell, Thaddeus. *A Renegade History of the United States.* Simon and Schuster, 2011.
Saliba, George. *Islamic Science and the Making of the European Renaissance.* MIT, 2007.
Shapin, Steven. *A Social History of Truth.* U. Chicago, 1994.
Shapin, Steven, and Simon Schaffer. *Leviathan and the Air-Pump.* Princeton, 2011.
Sivin, Nathan. "Shen Kua." In *Science in Ancient China.* Aldershot, 1995.
———. "Why the Scientific Revolution Did Not Take Place in China—or Didn't It?" *Chinese Science* 5 (1982): 45 (Revised 2005).
Sneader, Walter. *Drug Discovery: A History.* Wiley, 2005.
Sprat, Thomas. *The History of the Royal Society of London.* London, 1734 [1667].
Sun, Xiaochun. "State and Science: Scientific Innovations in Northern Song China, 960–1127." PhD thesis, U. Pennsylvania, 2007.
Thoren, Victor E. *The Lord of Uraniborg.* Cambridge, 1990.
Voelkel, James R. *The Composition of Kepler's Astronomia Nova.* Princeton, 2001.
Walsh, Judith E. *A Brief History of India.* 2nd ed. Facts on File, 2011.
Weinberg, Steven. *To Explain the World.* Harper, 2015.
Westman, Robert S. *The Copernican Question.* U. California, 2011.
Winchester, Simon. *The Man Who Loved China.* HarperCollins, 2008.

Wootton, David. *The Invention of Science*. HarperCollins, 2015.
Xu, Ting, and Khodadad Rezakhani. "Reorienting the Discovery Machine: Perspectives from China and Islamdom." *J. World Hist.* 23 (2012): 401.
Zuo, Ya. "Capricious Destiny: Shen Gua (1031–1085) and His Age." PhD thesis, Princeton, 2011.

AFTERWORD

Transistors: Gertner, 98–114; Riordan, 164–224. *IKEA:* Barthélemy; Collins; Kristoffersson, 15–21; Torekull, 49–84. *Amgen:* Binder, 26–27. *Interferon:* Edelhart; Jacobs; Pieters. **Quotations:** "*Helen spoke up,*" "*quail hunting,*" and "*small-town America*": Walton, 27, 41, 64. "*honest fight*": Torekull, 84. "*magic drug*": Edelhart.

Barthélemy, Jérôme. "The Experimental Roots of Revolutionary Vision." *MIT Sloan Mgmt. Rev.*, Oct. 2006.
Binder, Gordon M., and Philip Bashe. *Science Lessons*. Harvard, 2008.
Christensen, Clayton M. *The Innovator's Dilemma*. Harvard, 1997.
Christensen, Clayton M., et al. "What Is Disruptive Innovation?" *Harv. Bus. Rev.*, Dec. 2015.
Collins, Lauren. "House Perfect." *New Yorker*, Oct. 3, 2011.
Edelhart, Michael. "Putting Interferon to the Test." *NY Times Mag.* 130 (April 26, 1981): 32.
Galambos, Louis. "Theodore N. Vail and the Role of Innovation in the Modern Bell System." *Bus. Hist. Rev.* 66 (1992): 95.
Gertner, Jon. *The Idea Factory*. Penguin, 2012.
Goozner, Merrill. "The Longest Search: How Eugene Goldwasser and Epo Gave Birth to Biotech." *Pharm. Exec.* 24 (2004): 112.
Jacobs, Lawrence, and Kenneth P. Johnson. "A Brief History of the Use of Interferons as Treatment of Multiple Sclerosis." *Arch. Neur.* 51 (1994): 1245.
King, Andrew A., and Baljir Baatartogtokh. "How Useful Is the Theory of Disruptive Innovation?" *MIT Sloan Mgmt. Rev.*, Sep. 2015.
Kristoffersson, Sara. *Design by IKEA*. Bloomsbury, 2014.
Lepore, Jill. "What the Gospel of Innovation Gets Wrong." *New Yorker*, June 16, 2014.
Pieters, Toine. *Interferon*. Routledge, 2005.
Riordan, Michael, and Lillian Hoddeson. *Crystal Fire*. Norton, 1997.
Torkekull, Bertil. *The IKEA Story*. Trans. Joan Tate. Litopat, 2011 [1998].
Walton, Sam. *Made in America*. Doubleday, 1992.

ENDNOTES

INTRODUCTION

5 it was a piranha: Irreversible binders are sometimes used in laboratory experiments to explore the function of different proteins. Grabbing those proteins tightly makes it easier to detect their role in ordinary cell function.

8 unable to repeat its drug-discovery success: Global sales of Amgen and J&J's erythropoiesis-stimulating agents (Epogen, Aranesp, Procrit, Eprex) peaked at $9.8 billion in 2006. Between 1989 (the launch of Epogen) and 2004 (the launch of Sensipar), Amgen launched two follow-on, derivative products and a white-blood-cell stimulating agent (G-CSF) discovered at Sloan Kettering Cancer Center. Sources: Product revenues: SEC filings. G-CSF history: Welte. Goldwasser's role: Goozner; Goldwasser. Early Amgen: Binder; conversations with principals at Amgen and J&J.

9 still alive today, as I write this: As with any single case report of a new treatment, we can never know for sure how large a role our drug played in Alex's recovery. His cancer responded to the treatment (the tumors shrank significantly), but a subsequent large trial in melanoma for our drug did not succeed, and there were no subsequent clinical trials in Kaposi's sarcoma.

13 the steels used inside jet engines: The highest-strength steels are made of iron mixed with various transition metals (titanium, chromium, manganese, cobalt, nickel) and trace amounts of other elements. The science of adjusting tensile strength through small changes in structure is much more complex than the science of adjusting melting temperature. A material's melting point is mostly governed by the binding forces between molecules. Those binding forces are much stronger in iron than in water, which is why iron melts at 2,800 degrees Fahrenheit and ice melts at 32 degrees. A material's tensile strength, which is the amount of stress it can withstand before breaking, is very sensitive to a *different* element of structure: the arrangement of its atoms. Those arrangements and how they will affect fracture are difficult to predict, which is why the science of tensile strength is so complex.

CHAPTER 1

19 for the use of radar in battle: The term "RADAR" was coined later, in 1939, and commonly refers to devices that use a pulsed signal from the transmitter rather than a continuous signal, as described here. Although both depend on the reflection of radio waves, Young and Taylor's discovery would more accurately be described as detection by radio-wave interference (the "beat" method).

23 little interest in science: In his second inaugural address, FDR warned that "blindly selfish men" had turned science into a "ruthless master of mankind." Popular sentiment blamed labor-saving technologies for the high unemployment of the Depression. (FDR, Jan. 20, 1937.)

23 "among the minor miracles": Bush, who advised seven presidents, later described Hopkins as the greatest staff officer any president ever had: "I think that what attracted me most was his utter loyalty to his chief and his complete suppression of personal ambition." Hopkins had also been working on assembling a grassroots inventors council at the time they first met, so their ideas overlapped. See Kenny; Sherwood, 154; Bush, *Action*, 35.

26 an anonymous building at MIT: The group eventually included nearly two thousand people and nine future Nobel Prize winners.

27 would be portable: Radar is a variation on sonar, which had been developed during the First World War. Sonar detectors emit pulses of sound and listen for echoes. They can be useful over relatively short distances underwater, where there is little background noise, or in the air on dark, quiet nights (whales, dolphins, and bats all use sonar). Radar detectors emit pulses of light, rather than pulses of sound, and measure the reflected light coming back. Since light waves travel much farther than sound in air (you can see distant planes; you can't hear them), radar is better for long distances.

The light used in radar comes in different wavelengths, just like sound comes in different frequencies. Electric currents in larger antennas generate longer-wavelength light (radio spectrum); currents in smaller antennas generate shorter-wavelength light (microwave spectrum). Hence radio towers are tens of meters tall, while microwave antennas can fit in your hand.

It can be confusing that the term "radio" also refers to the device (now found mostly in living rooms of old movies) that converts radio-wavelength light into sound from a speaker. But a radio wave, as opposed to a radio device, just refers to light with a wavelength in the range of rougly one meter to one hundred kilometers.

27 British discovery of radar: In July 1934, at the Hotel New Yorker in Manhattan, Nikola Tesla, then 78 years old, announced what he considered the most important invention of his career: a method to send beams of particles through air that could destroy 10,000 enemy airplanes flying 250 miles away. The beam would "cause armies of millions to drop dead in their tracks," leaving no trace. The threat of annihilation would end all wars. The *New York Times* headline, naturally, announced a "death beam." No one in the US took it seriously.

Around the same time in the UK, Churchill had been warning of the rise of Germany, stating that London was "the greatest target in the world, a kind of tremendous fat cow . . . tied up to attract the beasts of prey." Churchill was in his wilderness years, out of government and often dismissed as a crank. He had somehow seen an internal memo by a physicist in the Air Ministry, Albert Rowe, stating that "unless science evolved some new method of aiding our defence, we were likely to lose the next war if it started within ten years." According to the physicist and radar historian David Fisher, Churchill called on the head of research at the Air Ministry, H. E. Wimperis, and insisted that he look into death rays. When Wimperis protested, Churchill loudly reminded him that the tank, which provided a crucial advantage in the First World War, had been dismissed by military planners at the start of that war, until he (Churchill) had rescued the idea. Shortly after the call from Churchill, Wimperis contacted Robert Watson-Watt, a radio engineer, "to advise on the practicability of proposals of the type colloquially called 'death ray.'" Watson-Watt and his assistant quickly established that death rays were impossible but that beams of electromagnetic radiation—light waves—might be used for detection. In February 1935, an Air Ministry committee created a small team to investigate Watson-Watt's idea. Four years later the Chain Home radar system was the result.

And that's how death rays saved England.

(See Fisher, *Summer*, 54–68; Churchill, 1934; and the additional references in the bibliography on radar in England.)

27 concentrate their limited forces: Also critical were the development of new mathematical techniques and a sophisticated, real-time data management system (the Dowding system) to process the data from radar. Those techniques gave rise to what is now called operations research (Budiansky; Hartcup, 100–21).

28 jump-started Loomis's efforts: On September 28, 1940, the British delivered to Loomis a palm-sized power generator needed to create portable microwave radar. That device, called a cavity magnetron, was described by an American military historian as "the most valuable cargo ever brought to our shores" (Baxter, 142; Conant, 179–208; Phelps).

29 Britain was running on fumes: Years later Churchill wrote that "the Battle of the Atlantic was the dominating factor all through the war. Never for one moment could we forget that everything happening elsewhere, on land, at sea, or in the air, depended ultimately on its outcome. . . . The only thing that ever really frightened me during the war was the U-boat peril" (*Ring*, 6; *Finest Hour*, 529). Roosevelt agreed. In a May 1941 cable to Churchill, he wrote that the war would be won or lost in the Atlantic.

29 German codebreakers: The story of British scientists breaking the German Enigma codes, the Ultra program, has been well told. Ultra, however, had little impact on the Battle of the Atlantic. This was primarily due to the even greater success by the Germans in cracking British codes (a story that has not been well told): German intelligence deciphered a high fraction of the most critical Allied naval messages from the summer of 1938 through the end of 1943. In a long-suppressed, confidential postwar analysis, the horrified commander of British signals intelligence noted, "This deplorable record of enemy achievements is substantiated beyond a doubt by (a) interrogation of high German Naval Officers . . . and (b) examination of the actual German Logs containing our deciphered signals." See Tighe for the analysis and Syrett, 96–180, for a vivid description of how signals intelligence played out in real time. Summaries of British vs. German signals intelligence: Erskine; Gardner, 210–18; P. Kennedy, 23, 35, 61–63.

30 a pilot could calculate his location: The Allies built antenna stations along the US and Canadian coasts, as well as those of Greenland and Iceland, which provided coverage across the Atlantic. The system was initially called LRN for Loomis Radio Navigation and then changed to LORAN, for LOng-RAnge Navigation, at Loomis's request. LORAN allowed planes and ships to determine their location within 1 percent accuracy up to 1,400 miles from a station. LORAN was widely used until the 1990s, when it was replaced by GPS. See Baxter, 150–52; Conant, 231–34, 265–67.

31 U-boats were unable: German codebreakers, intercepting radio traffic, were stunned to discover that no more than one or two Allied planes were protecting the convoy. In his war diary, Doenitz concluded, "The enemy's radar hardly missed a boat" (Syrett, 134).

32 "won the Battle of the Bulge": Artillery shells previously used timed fuses. After estimating by eye the time of flight to a target, a gunner would fire his weapon and hope the fuse would go off somewhere near the target, which was especially difficult for moving targets. The radar-timed fuses, called proximity or VT (variable-timed) fuses, eliminated that guesswork, which dramatically improved firing efficiencies. The prox fuses transformed the ability of ships and bases to protect themselves against incoming aircraft and provided much more devastating artillery firepower on land. Shortly after the Battle of the Bulge, one US officer recorded, "PW [prisoner-of-war] reports are unanimous in characterizing our artillery fire as the most demoralizing and destructive ever encountered" (Baldwin, 280).

33 Einstein's famous letter: On October 11, 1939, economist Alexander Sachs brought FDR a letter from Albert Einstein. Known as the Einstein-Szilárd letter, it warned of recent work suggesting that "the element uranium may be turned into a new and important source of energy" and that "extremely powerful bombs of a new type may thus be constructed."

33 still controversial eight decades later: For a detailed description of Bush's role in launching the nuclear program, see Goldberg, "Bush and the Decision."

The Manhattan Project achieved the first controlled nuclear explosion on July 16, 1945, two months after Germany surrendered. The first bomb detonated over Hiroshima on August 6; the second over Nagasaki on August 9. Japan surrendered shortly afterward.

Essays by Stimson and others published shortly after the war asserted that use of nuclear weapons ended the war with Japan sooner and saved up to a million American lives. Those essays were widely read and accepted by the public. At the time, however, many prominent military leaders publicly disagreed with the official view. (General Curtis LeMay, for example, who oversaw the Japanese bombing raids, stated, "The atomic bomb had nothing to do with the end of the war." Every major Japanese city had already been decimated by Allied bombers; the country was embargoed; its navy was finished; oil and food supplies were nearly gone; and Japan's only ally had surrendered.)

The historian Sam Walker has said recently that the decision to use the bomb against Japan "has been, in terms of longevity and bitterness, the most controversial issue in American history." The controversy centers around the reasons behind the Japanese surrender (use of the bomb vs. Russia's declaration of war on Japan on August 8, which ended its last hope of a mediated surrender); the accuracy of the postwar justifications (which historians on all sides have concluded were fabricated); and Truman's motivations. For an excellent recent history, which incorporates sources only available since the 1989 death of Japan's wartime emperor and the 1991 collapse of the Soviet Union, see Hasegawa, *Racing the Enemy*. For balanced recent summaries of the debate, see Walker, *Destruction*; and the essays in Kelly, 319–422, and in Hasegawa, *Pacific*. LeMay: Bernstein.

Nearly all historians agree that records show little discussion among Truman and his small circle of advisors about the merits of using the bomb once it was ready. The decision to bomb cities densely populated with civilians had been made years earlier by both the Allied and Axis powers. (More civilians died in the firebombing of Tokyo in March than in either Hiroshima or Nagasaki in August.)

Churchill may have articulated the view at the time most clearly when he told an anxious Niels Bohr, in 1944, that there was no need to worry about a postwar nuclear world: "After all, this new bomb is just going to be bigger than our present bombs and involves no difference in the principles of war" (Jones, "Churchill," 88). Only in hindsight did this view change.

35 "epoch-making," trillions: At a conference marking the fiftieth anniversary of *Endless Frontier*, one historian noted that the report has risen to "biblical status" in science policy circles, widely scrutinized and interpreted, often with contradictory conclusions. The National Science Foundation, the National Institutes of Health, and many other research agencies are modeled on the principles described in Bush's report. For more on the immediate aftermath, see chapter 8. For accounts of the long-term impact of *Endless Frontier*, see CSPO, 1–35; England, 3–110; Greenberg, 68–148; Kevles, 267–321; Zachary, 240–60. For recent reviews of the economic impact of federal science policy, see Lane; Jarboe; PCAST; and NAS.

37 residue of design: The saying is attributed to Branch Rickey, the Hall of Fame baseball executive, who created baseball's farm league system (the Major and Minor leagues); built eight World Series teams; and found, signed, and started Jackie Robinson, the first African American baseball player (Breslin, 73).

39 Avastin . . . James Bond and *Star Wars*: For more on movies and drugs, including the Bond-Connery loonshot, see chapter 5. The full title of the 1976 fourth draft of the *Star Wars* script, used when filming began, was "The Adventures of Luke Starkiller as Taken from the 'Journal of the Whills.'" The rejected initial treatment in 1973 was called "The Star Wars" (Rinzler).

For more on loonshot nurseries within *industries* (drug discovery; film) rather than *companies*, see chapter 9. That chapter also describes an additional principle necessary for a successful loonshot nursery, which would be getting too far off topic for this chapter: the principle of critical mass.

40 "associating with military men": Bush continued: "Military men learn the art of command; it is central to their whole professional careers. They also learn to behave well in tight groups . . . it is an incorrigible officer indeed who does not emerge with an exceedingly attractive attitude of courtesy in places where courtesy is called for" (Bush, *Action*, 298).

41 equal-opportunity respect: In biotech, creating a new drug is so complex that it requires massive teams of both artists and soldiers: biologists, chemists, physicians, marketers, regulatory specialists. The groups often distrust each other. Biologists may see chemistry as more magic than science, medicine as not a science at all, and businessmen as aliens from an evil planet. Chemists tend to view themselves as the only true drug developers. Medics may view themselves as the only ones who matter in the end; enough said. The businessmen think of themselves as calm caretakers in a lunatic asylum. Getting a drug approved and distributed to patients requires all these groups to cooperate. Anyone managing the effort has to learn to overcome the distrust between groups. Which begins with overcoming their personal preferences for their own kind. In other words, succeeding in biotech, just as Bush showed with the military, and Vail and Jobs showed with technology companies, requires learning and practicing equal-opportunity respect.

41 stayed out of the details of any one loonshot: By contrast, Bush's counterpart in Great Britain took the opposite approach. Frederick Lindemann, Churchill's science advisor, argued passionately for his idea of using floating mines in the sky to defend against enemy aircraft. His political antics dangerously delayed the British radar program. Churchill was also unable to resist diving deep—very deep—into loonshots. He insisted, for example, on a secret project to build a two-million-ton floating island made of ice to carry aircraft. He provided specific design instructions, including what sort of ice to choose and how to spray the ice. The idea made its way to Roosevelt, who asked Bush about it. Bush's brief answer: one could build an aircraft carrier for that cost and "it would not melt." FDR dropped the subject. (Snow, 10–38; Bush, *Action*, 123–25; Perutz.)

CHAPTER 2

48 low-fat, high-carbohydrate diets: The recommendations were controversial at the time. No controlled studies had shown that reducing dietary fat improved health, prompting the president of the National Academy of Sciences to state at a congressional hearing, "What right has the federal government to propose that the American people conduct a vast nutritional experiment, with themselves as subjects, on the strength of so very little evidence?" Subsequent large clinical studies repeatedly failed to find any evidence supporting health benefits to low-fat diets. The low-fat guidelines, however, persisted until only very recently. See Taubes, 3–88; NRC, 10.

52 the drug worked: In a recent email exchange, Dr. Yamamoto noted that patient S.S. was eventually cured, at age 23, by a combination of statin treatment, two coronary bypass surgeries, and plasma apheresis, which enabled her pregnancy at age 26. It was the consistent effect of treatment with Endo's statin at low doses on a series of subsequent patients with heterozygous (more moderate) FH that convinced Drs. Yamamoto and Endo and other researchers that the drug was effective.

53 by just four atoms: The Merck compound (referred to as MK-803, lovastatin, mevinolin, or Mevacor) and Endo's compound (referred to as ML-236B, compactin, or mevastatin) are identical except for one hydrogen atom off a side ring in mevastatin, which is replaced by a methyl group (one carbon atom and three hydrogens) in lovastatin. See, for example, Alberts, "Lovastatin."

54 to collaborate, rather than compete: Other accounts by Merck scientists: Alberts; Cordes; Tobert; Vagelos. Letters documenting the exchange between Merck and Sankyo, from April 1976 through October 1978, were kindly provided to the author by Akira Endo. Merck–Sankyo letters were primarily exchanged between H. Boyd Woodruff (executive administrator, Merck Research Labs) and Dr. Issei Iwai (director, Product Planning Dept., Sankyo), although one was addressed to Endo directly and one to Dr. Ko Arima directly (Endo's supervisor, the head of Sankyo Central

Research Labs). On April 16, 1976, Woodruff wrote Iwai: "The properties of the compound [ML-236B] are very interesting and the biochemists in our laboratory would like to evaluate it. . . . We hope that as a result of these exchanges, a product will be found which is suitable for license and royalty return." A letter from Merck dated Sep. 23, 1977, summarized their work together: "Your compound ML-236B has remarkable properties. . . . It seems evident that a practical therapeutic application will develop from Dr. Endo's research program."

In his memoir, Vagelos, who joined Merck in 1975, describes Sankyo as "dogging our steps" (p. 137). Endo discovered the first statin in 1973; filed a patent on behalf of Sankyo in June 1974; demonstrated statin activity in animal models in early 1976; disclosed proprietary data to Merck from 1976 to 1978 at Merck's request; and initiated the first human clinical trials in 1978, demonstrating that statins can help patients. All this took place before Merck began its statin program, which was, according to Merck, in October 1978.

54 terminated Merck's program: Vagelos describes repeatedly asking Sankyo in 1980 for their dog study results and being surprised that they rejected his requests. "I thought this sort of direct inquiry might be successful because Merck & Co., Inc., had strong ties to the Japanese pharmaceutical industry," he writes. Vagelos describes Sankyo's lack of response as "an ethical issue" (Vagelos, 149–50).

55 "such a dramatic effect": For excellent histories of the cholesterol controversy and statins, see Steinberg; Goldstein. For recent reviews of the benefits, risks, and impact of statins, see Goldfine; Collins.

56 have exceeded $300 billion: The Merck franchise includes sales from Mevacor (launched in 1987); an improved version, Zocor (launched in 1990); and Vytorin (a combination product: Zocor combined with Schering-Plough's Zetia, launched in 2004). Net income from Vytorin was split with Schering-Plough. Figures are from company reports. The two other leading statins to date have been Lipitor and Crestor. Cumulative sales for Lipitor (developed by scientists at Warner-Lambert, now marketed by Pfizer) have exceeded $140 billion. Cumulative sales for Crestor (developed by scientists at Shionogi, now marketed by Astra-Zeneca) have exceeded $50 billion.

59 increased by $38 billion: The change in market value of Genentech between the first announcement of positive results in colon cancer on May 19, 2003, and the FDA approval on February 26, 2004, was $38 billion. The one-day change from the announcement on May 19 was $9 billion.

In 2006, a large clinical trial of a derivative of Avastin called Lucentis showed that it can reverse a form of blindness. (Avastin injections have been shown to provide similar benefits.) The accompanying editorial in the *New England Journal of Medicine* described the results as "miraculous," a term the *NEJM* has applied to trial outcomes on only one other occasion over the past twenty years (gastric bypass surgery). See Stone; Rosenfeld.

59 its share of $300 billion: Because of some overlap between its patents in the US and patents that Endo and Sankyo had filed in Japan, Merck was eventually forced to license certain territorial rights for Mevacor from Sankyo.

CHAPTER 3

66 Apple, Microsoft, and GE: AT&T's peak share of total US stock market value reached 13 percent (1932); Apple, Microsoft, and GE each peaked at less than 4 percent of total US market value. Data: U. Chicago Center for Research in Security Prices, US Stock Database (Nov. 2017).

69 Crando: For those too young to remember a pinnacle of the Sylvester Stallone oeuvre—the reference is to *Rambo*.

73 impossible to copy: Sabre was eventually spun out as an independent company and is no longer part of American.

83 passed the plan on to Truman: Truman sent General Albert Wedemeyer to China to investigate further. Despite the general's consistent account and his recommendation to increase aid to Chiang Kai-shek, Truman declined, acting on the advice of his secretary of state, George Marshall. In 1949, Mao defeated Chiang and took control of China. Chiang and the Nationalists fled to Taiwan. The outcome led to the acrimonious "who lost China" debates in the US in the 1950s (Wedemeyer; May).

84 the first ballistic missile: The Me 262 was called the *Schwalbe*, German for "swallow." Hitler didn't like the name, so it was changed later to *Sturmvogel*, "Storm Bird."

In ordinary engines, boiling water or exploding gas fires pistons back and forth inside a cylinder, rotating a lathe or axle or propeller. All planes prior to World War II flew with piston engines and propellers. In a jet engine, the exhaust from a controlled explosion of fuel creates the forward thrust. An "airbreathing" jet, which mixes the fuel with air, powers jet aircraft. Rockets take in no air; a chemical mix inside the rocket combusts to produce the exhaust.

84 dismissed by academics and the military: Goddard died from cancer in August 1945, too soon to see the US put his ideas into practice, but just long enough to examine captured German technology and recognize it as his own.

After World War II, the US recruited the German rocket scientists who had studied Goddard's work to help build the US space program. In 1959, NASA named its largest space flight research center after Goddard. In 1960, the government admitted its rocket program infringed on Goddard's original patents and awarded his estate $1 million.

Vannevar Bush missed the potential of jet engines, his most serious oversight of the war. The V-2 traveled over two thousand miles per hour, untouchable by antiaircraft fire, too fast for interception by any plane. Fortunately for the Allies, the German jets and rockets arrived too late to make a difference to the outcome of the war. They were neutralized by the Allies' overwhelming air superiority by 1944, which allowed the Allies to bomb the runways used by the jets and the launch sites used by the rockets (Boyne; Bush, *Arms,* 71–89; Clary; Pavlec; King).

85 "he went from Jesus to Judas": Several recent biographers have written extensively on the controversy surrounding the portrayal of Lindbergh as a Nazi sympathizer and anti-Semite. Defenders note that Lindbergh visited Germany in the 1930s at the request of the US State Department and military; reported to US and British political and military leaders, at their request, on German air force capabilities; and that his assessment of the Luftwaffe's strength was crucial in mobilizing US and British forces. Defenders also note that Lindbergh's antiwar views at the time were in line with a majority of Americans; that they were motivated by his view of Stalin as the greater threat than Hitler; and that his views on the Nazi regime, like many others at the time, changed after reports emerged of the massive pogrom, *Kristallnacht,* organized by the Nazis against the Jews late in 1938. (Lindbergh abandoned plans to move to Germany, writing that he "did not wish to make a move which would seem to support the German action in regard to the Jews.") Others note Lindbergh's sympathy during the prewar period to the extreme views on race and eugenics espoused by a number of his friends and mentors (Alex Carrel, Truman Smith, Henry Ford), and argue that his flawed assessment of German air force strength was critical to the Munich appeasement pact in 1938.

Nearly all biographers agree that Lindbergh was politically naïve and that he allowed his fame to be used by political leaders on all sides for their own agendas. (Even the senior Nazi officer Albert Speer, years after the war, described Lindbergh as "naïve.") Lindbergh's choice to refrain from commenting publicly on the campaign against him, despite factually inaccurate attacks, further eroded his public image. After Pearl Harbor, Lindbergh turned strongly in favor of the war, but his public image never recovered. (See Berg, 355–458; Olson; Wallace. Speer cited in Wallace, 193; Lindbergh cited in Berg, 380.)

85 have no connection with Lindbergh: Lindbergh eventually found consulting contracts with two aircraft manufacturers and talked his way into a Marine air squadron based in the Pacific in 1944 as an industry consultant, ostensibly to evaluate aircraft performance. As a civilian, Lindbergh flew 50 combat missions against the Japanese, which was technically illegal. One pilot remembered, "Lindbergh was indefatigable. He flew more missions than was normally expected of a regular combat pilot. He dive-bombed enemy positions, sank barges and patrolled our landing forces on Noemfoor Island. He was shot at by almost every anti-aircraft gun . . . in western New Guinea." Lindbergh taught pilots his long-distance flying techniques: by lowering the revolutions per minute of their engines and using more "boost" (manifold pressure), they could preserve fuel and extend their flight time and combat radius by as much as 50 percent. The distance improved the safety of the planes and allowed the squad to surprise the enemy far deeper into enemy territory. General MacArthur heard about it, sent for Lindbergh, and told him the technique was a "gift from heaven." He asked Lindbergh to teach other squadrons and gave him permission to fly any plane he wanted. (One pilot remembered: MacDonald; MacArthur cited in Berg, 452.)

CHAPTER 4

95 can't detect polarization: To visualize polarization, imagine fastening one end of a rope (light beam) to a wall at the height of your hip, then backing away from the wall holding the other end until the rope is taut. Jiggling the rope up and down creates a vertically polarized wave. Jiggling the rope left and right creates a horizontally polarized wave. Light beams are propagating vibrations of electric and magnetic fields. The motion of the rope corresponds to the electric field oscillations.

96 Polarizing filters function: The analogy with a drone is not perfect, since light acts like a wave. Light polarized at 45 degrees is an equal mix of waves polarized horizontally and vertically. A horizontally polarized filter, more accurately, picks out just the horizontal portion of the wave.

98 could never convince them: Many discussions blame auto manufacturers for failing to adopt the idea because of the additional expense. Coating windshields with a polarizer, however, could reduce all visibility by up to 50 percent. The reduction in visibility is a serious safety concern, especially in low-light conditions.

100 more than two million pixels: Because light from LCD screens is polarized, it can be blocked by polarized filters. (You can test this by holding polarized sunglasses in front of a screen and rotating the glasses through ninety degrees.) The difficulties of reading LCD displays through polarized sunglasses have reduced the use of those sunglasses.

110 chemistry of color development and film transparency: Photographic film uses *subtractive* color: chemicals in the film store a color's opposite (cyan for red; magenta for green; yellow for blue). Color transparencies use *additive* color, storing the original color.

113 digital cameras using CCDs: CCD chips are actually analog, not digital, devices. The sensors produce a *continuous* (analog), rather than *discrete* (digital), voltage corresponding to the intensity of light hitting a pixel, just like a bucket collecting rain measures a continuous, rather than discrete, water level. Analog-to-digital converters were eventually added to CCDs so their output could be stored on digital memory chips. Although other terms for CCD devices used at the time ("electro-optical imaging" or "solid-state devices") are more accurate, the term "digital photography" has come to distinguish a photoelectric from a photochemical process. I'm using the term in the common, current sense.

113 He was quickly selected: Killian was a friend and colleague of Land's, and eventually a Polaroid board member. The panel that Land chaired was called initially the Intelligence Panel and then, at various times over the next twenty years, the Land Panel or the Land Reconnaissance Panel. Killian and Land worked closely together and generally met together with Eisenhower, and subsequently with Kennedy, Johnson, and Nixon. (Killian chaired the umbrella group, the Technological Capabilities Panel.)

114 "his second term in office": The NRO official historian, writing in 2012, noted that the agency had never before, or since, received a timeline referring to election cycles (Perry, 526).

115 the other side: Edwin Land: The interagency battle played out as the Air Force (film scanners) vs. the CIA and Land (digital). Land developed the proposal of using digital sensors and helped create and guide the Directorate of Science and Technology (DST) within the CIA, which formally backed the digital proposal; he also presented the idea to the president. In a confidential memo, the NRO director noted, "If EOI [electro-optical imaging] is a technology-driven development, Dr. Land is the main driver" (Perry, 527). For Land's role in developing the DST within the CIA, as well as protecting it when Kennedy wanted to decimate the CIA following the Bay of Pigs, see Richelson, 67–72.

117–18 Pacific Southwest Airlines: PSA, which operated from 1949 until it was acquired by USAir in 1986, was the first major discount carrier, initially flying routes in California only. The noses of its planes were painted to look like a smiley face (the company slogan was "Catch our smile"). It was the model for today's surviving Southwest Airlines, which began operations in 1971. Lamar Muse, the founding president of Southwest, noted, "We don't mind being copycats of an operation like that" (*Business Week*, "Love").

119 before the wheel stops turning: Physicists call a state in which the phase transition is temporarily prevented, such as supercooled water, "metastable."

CHAPTER 5

125 the demoralizing attacks: A page 1 *Wall Street Journal* story quoted Wozniak saying the Apple II "had been ignored in the hope that it will die and go away" (Bellew; Feb. 7, 1985).

126 launched Newton on the path: Newton had described initial thoughts about gravity and planetary motion a decade earlier in his notebooks from 1666 to 1668, but in the context of Descartes's "vortex" theory. He had abandoned mechanics and gravity and was studying alchemy when Hooke reached out to him in 1679. Hooke suggested the essential idea that planetary motion should be decomposed into a linear inertial component and a centripetal attractive force directed toward the sun. That idea is the starting point, Proposition 1, of Newton's *Principia.* (One historian describes Newton's attempts to revise the priority date of his theory and deny Hooke's contribution as "bogus history"; another as a "fairy-tale.") Later, as president of the Royal Society, Newton attempted to write Hooke out of history, which he nearly achieved. Hooke's role has only been rediscovered and assessed by historians over the past few decades.

For more on Kepler, see chapter 9. For good summaries of the Hooke-Newton controversy, see Cohen, 223–79; Gal, 161–230; Jardine, 1–19; and Nauenberg. For Newton and calculus: "Foreshadowings of the principles and even of the language of [the infinitesmal] calculus can be found in the writings of Napier, Kepler, Cavalieri, Pascal, Fermat, Wallis, and Barrow. It was Newton's good luck to come at a time when everything was ripe for the discovery, and his ability enabled him to construct almost at once a complete calculus" (Ball, 347). For Newton's other predecessors see Hall; Whiteside. Hooke's wings, bouncy shoes, and marijuana experiments: Inwood, 21, 334, 398. Bogus history: Cohen, 248; fairy-tale: Whiteside, 14. Although accounts of the Hooke-Newton controversy often focus on the priority in deriving the familiar inverse-square form of gravity, many had combined Kepler's period law with Huygens's centrifugal force to derive the same thing (a one-line step).

126 "yet is he very merry": From *An Account of the Plant call'd Bangue [Gange by the Moors], before the Royal Society, Dec. 18. 1689* (Hooke, 210): "It is a certain plant which grows very common in India . . . This Powder being chewed and swallowed, or washed down, by a small Cup of Water, doth, in a short Time, quite take away the Memory and Understanding; so that the Patient understands not, nor remembereth any Thing that he seeth, heareth, or doth, in that Extasie, but becomes, as it were, a mere Natural, being unable to speak a Word of Sense; yet is he very merry, and laughs, and sings, and speaks Words without any Coherence, not knowing what he saith or doth; yet is he not giddy, or drunk, but walks and dances, and sheweth many odd Tricks; after a little Time he falls asleep, and sleepeth very soundly and quietly; and when he wakes, he finds himself mightily refresh'd, and exceeding hungry."

127 or humanize deities: Edmond Halley wrote the foreword (opening inscription) to the *Principia*:

O you who rejoice in feeding on the nectar of the gods in heaven
Join me in singing the praises of NEWTON . . .
No closer to the gods can any mortal rise.

Voltaire wrote, "The catechism reveals God to children, but Newton has revealed him to the sages!"

136 convinced Disney to purchase a handful of PICs: In what might be called an uncredited cameo, it was Catmull and Smith's work with Schure and his *Tubby the Tuba* animators that helped convince Disney (Price, 93).

137 partnership with a large pharma company: The partnership was with Eli Lilly. The goal was a synthetic version of human insulin for treating diabetes. For the previous half century, Lilly and other suppliers had to grind up pancreases from pigs or cows to extract insulin. Genetic engineering allowed insulin to be grown in a lab. See Hughes, 75–106.

144 $10 billion in annual sales: Revenues based on US sales only (excluding royalties on ex-US sales) in the last full year before Genentech was acquired by Roche (Genentech 2008 annual report).

INTERLUDE

154 both *emergent properties*: In *The Self-Organizing Economy*, which describes connections between economics and the science of emergence, Paul Krugman, a Nobel laureate in economics, noted, "When Adam Smith wrote of the way that markets lead their participants, 'as if by an invisible hand,' to outcomes that nobody intended, what was he describing but an emergent property?"

Reacting to Greenspan's comment, Krugman wrote in the *New York Times* that he was "left speechless" by Greenspan's lack of self-awareness of his role in causing the crisis. "Alan Greenspan continues his efforts to cement his reputation as the worst ex-Fed chairman in history" (March 30, 2011).

154 explode into a wildfire: The story of why markets will always crash is a little more complicated, but it involves the same underlying principle of two competing forces.

155 prized his work on ethics: Smith wrote, "All for ourselves and nothing for other people is a vile maxim" (cited in McLean, ix). For more on the misinterpretations of Smith, see Kennedy, 251–59; McLean, viii–ix, 82–98; Rothschild, 2–5, 116–56; Wight.

On Smith's preference, from a contemporary: "One ought not, perhaps, to be very much surprised that the public does not do justice to the works of A. Smith, since he did not do justice to them himself, but always considered his *Theory of Moral Sentiments* as a much superior work to his *Wealth of Nations*" (Sir Samuel Romilly, Letter to Madam G—, August 20, 1790). Romilly was bemoaning "how little impression his [Smith's] death has made here. Scarce any notice has been taken of it, while for above a year together, after the death of Dr. Johnson, nothing was to be heard of but panegyrics of him."

156 attributed that meaning retroactively to Smith: Ovid, Shakespeare, Voltaire, and Defoe used the phrase, as did many contemporary writers. Smith taught rhetoric before he became a professor of philosophy and lectured on Shakespeare's use of metaphor; he would have been well aware of the usage (his library contained many books using the phrase). The economist Gavin Kennedy notes, "If Samuelson [the author of the textbook] had read *Moral Sentiments* and *Wealth of Nations* for himself through its many editions and translations well into the 1970s, instead of recalling what he was taught at Chicago by his tutors and then passing on the same error to hundreds of thousands of readers of Economics, many of whom became tutors themselves, the current epidemic of misleading ideas about invisible hands may have become containable." See also Rothschild, 2–5, 116–56; Wight, "Smith."

157 whether bakers sell cupcakes or bread: So many have written on Smith's economics. So few have written on his alliteration. "It is not from the benevolence of the butcher, the brewer, or the baker that we expect our dinner, but from their regard to their own interest" (Smith, *Wealth*, 26).

157 economists have aspired: For a history of the interplay between physics and economics, see Mirowski, especially chapter 7: "The Ironies of Physics Envy." For Newton and Smith: Montes; Hetherington. Newton was much less dogmatic about fundamental laws than many of his disciples. He considered his gravitational law an approximation until something better came along (Montes, 41–42; Schofield, 177: "Newton was not a Newtonian").

157 by the forces of history: Full quote: "That the essential role played by higher organizing principles in determining emergent behavior continues to be disavowed by so many physical scientists is a poignant comment on the nature of modern science. To solid-state physicists and chemists, who are schooled in quantum mechanics and deal with it every day . . . the existence of these principles is so obvious that it is a cliché not discussed in polite company. However, to other kinds of scientist[s] the idea is considered dangerous and ludicrous, for it is fundamentally at odds with the reductionist beliefs central to much of physics. But the safety that comes from acknowledging only the facts one likes is fundamentally incompatible with science. Sooner or later it must be swept away by the forces of history." (Laughlin, "Theory," 30.)

CHAPTER 6

166 Over a dozen Nobel Prizes: In certain metals, all electrical resistance suddenly vanishes below a very low temperature—ordinary metallic friction just disappears. The sudden change marks the transition from an ordinary metal to a *superconductor*. Albert Einstein, Niels Bohr, Werner Heisenberg, and Richard Feynman invented the theory of relativity, quantum mechanics, and particle physics as practiced today. All tried and failed to explain superconductivity. The mystery of superconductivity remained unsolved for 46 years, from its discovery in 1911 until 1957, when a trio of physicists showed that below a critical threshold temperature, electrons inside a metal will pair up: as if lone individuals meandering around a dance floor suddenly hear music and rush to find a partner. Phil Anderson, mentioned earlier, showed that symmetry-breaking associated with those electron pairs explains why electrical resistance falls to exactly zero.

A handful of particle physicists applied Anderson's ideas to solve another long-standing mystery: how we should think about the origin of mass in the universe. They jointly came up with the idea for what is now called the Higgs boson. (Murray Gell-Mann, who coined the term "quark" and helped create the standard model of particle physics, for which he earned the 1969 Nobel Prize, has argued it should be called the Anderson-Higgs boson.)

Histories of superconductivity: Schmalian; Cooper. Popular histories of the Higgs search: Carroll, 135–62; Gell-Mann, 193–94; and Lederman. Technical histories of the Anderson-Higgs mechanism: Brown; Hoddeson, 478–522; Anderson, 4–49, 115–19; Witten.

171 the epidemic phase transition: In the late 1970s and early 1980s, mathematicians formally proved the equivalence between models of percolation and models of the spread of disease.

173 pushed the forest across the dashed line: More accurately, it is the ratio between the sparking rate and the tree-regrowth rate that matters. When the sparking rate is low compared to the regrowth rate, the density of trees in a forest gradually increases, until it crosses the contagion threshold. For fire fact-hounds: most fires in the United States are caused by people. A 2002 study of 538,809 wildfires in the US from 1970 to 2000 found that 57 percent of fires were caused by people vs. 43 percent by natural events, which were mostly lightning (Brown, 15).

174 Hundred-acre fires should occur one-tenth as often: The ratios are the simplest example of a power law distribution, where the frequency varies in exact inverse proportion to size. Sophisticated forest-fire models predict an exponent closer to 1.15 than 1.0. For reviews, see Hantson; Zinck.

176 been in a movie with Bacon: According to the Oracle of Bacon site, as of October 2018, out of 2.9 million actors in their database, a total of 2.3 million actors have a link to Bacon. Of those, 3,452 (0.1 percent) are one degree removed; 403,920 (17 percent), two degrees; and 1,504,560 (64 percent), three degrees.

177 cited more than Einstein's paper on relativity: The Watts-Strogatz 1998 paper is followed closely by the Barabási-Alberts 1999 paper, which proposed a similar concept, adding the idea of "preferential attachment": nodes with more links get friended more. In other words, popular kids get liked more. (The same principle underlies Google's PageRank search algorithm.) According to the curated list maintained by the high-energy physics database INSPIRE, the two highest-cited papers in "fundamental" physics (excluding materials science and calculational techniques) are Steven Weinberg's 1967 paper on the standard model of particle physics (5,905 citations) and Juan Maldacena's 1999 paper on string theory (4,651 citations). Citations do not necessarily reflect importance; Einstein's papers, for example, are rarely cited now because the ideas have been so integrated. All citation numbers are from Web of Science Core Collection.

180 by an unusual power: 2.5: Four raised to the power of 2.5 is 32. Because trees in a forest are limited to infecting only trees in close physical proximity but people can spread information quickly to a large group, the form of the power laws is different in the two cases. For experts: an exponent of 2.5 is what you get from percolation theory when the number of neighbors grows large (an infinite-dimensional network).

184 other forms of violent conflict: For more on the application of techniques from statistical physics to network science and human conflicts, see the references in the source notes.

CHAPTER 7

187 anyone could have spiritual visions: "Jesus's claim to be the Christ or Messiah was never meant to be exclusionary, Emerson asserts," wrote Richard Brodhead in 2003. "The message of the living Jesus was just the opposite. . . . He invites [his followers] not to the role of minister, holder of an official position in an institutional church, but rather to the role of preacher-prophet: a proud enjoyer of access to the divine who awakens others to their own comparable powers." (Brodhead, 56–57.)

191 roughly one in ten: In drug discovery, the probability that a drug entering clinical trials will make it all the way through to FDA approval has consistently averaged, across many studies, around 10 percent. (It varies from one in five to one in 20 depending on the disease area and type of drug.)

In the film industry, a widely quoted statistic states that less than one in five released films will earn a positive return. Unlike drug discovery, however, film production is not heavily regulated, and therefore data are nearly impossible to confirm. Ancillary sales (e.g., streaming video) can increase that rate. Despite some benefit from ancillary sales, however, the 20 percent rule measures only the percentage of *released* films. Far more films are completed than are released: thousands of films are produced each year; only a few hundred make it to theaters. The very low ratio of film projects that start to film projects that finish, as with drugs, means that the likelihood an investor in an early-stage film project will earn a positive return is extremely low. (Drug discovery: Wong. Film: Sparviero; Epstein.)

197 Your stake in that future income grows: These examples refer to equity in the familiar sense of overall stake in the company's success (e.g., stock or stock options or profit share). Equity can also be tied directly to project success, separate from company success, which is discussed later.

200 skill and politics ratios: If the ratio of project–skill fit to return on politics (*F*) gets too far out of whack, organizations tend to rearrange so the ratio comes back closer to one. For example, employees who breeze through their current projects—a very high skill ratio—may be promoted until their span of projects is more challenging.

CHAPTER 8

207 since Nikola Tesla made headlines: see chapter 1 note for *"British discovery of radar"* on page 317.

213 benefit of future job offers: Some companies hide their best employees for just that reason—fear that star talent will be poached. Those companies will never attract and retain great talent. Stars will find their way to a competitor that is unafraid of letting them shine in front of their peers.

213 patient groups: The Leukemia and Lymphoma Society, the Multiple Myeloma Foundation, and the Cystic Fibrosis Foundation—all patient-led advocacy groups—have formed effective partnerships with biotech companies that have resulted in important new drugs. Vertex Pharmaceuticals' two breakthrough drugs for treating CF, for example, were developed with the close support of the CF Foundation.

219 a big *G* is a bad thing: Bloom; Wade. (Quotation: Wade, 528.) The "tournament theory" model in economics, in which employee pay is determined by relative rankings, might make sense in an imaginary world where everyone works on a project on their own. In the real world, it will exacerbate the problem described in this chapter: it will inflame the battles that destroy cohesiveness and encourage politics. Nurturing fragile loonshots requires individuals to unite around a big, exciting, common goal rather than compete to destroy each other.

222 leads to fewer failures: For an analysis of these two choices by economists, see Sah and Stiglitz; Csaszar.

225 jail terms: For more on behavioral economics, see *Thinking, Fast and Slow* by Daniel Kahneman (from which the jail term example is drawn, pages 225–26); the *Pre-*

dictably Irrational series by Dan Ariely; or the *Freakonomics* collection and blog by Steven Levitt and Stephen Dubner. For a recent summary and entertaining history, see *Misbehaving: The Making of Behavioral Economics* by Richard Thaler.

225 for both types of deliveries: See Allin for a recent economic analysis, and NPW for an assessment of likely reasons and common myths behind the steep rise in C-section rates. Although legal pressures have been frequently cited as contributing to the rise, recent studies have shown they have played little role (Sakala).

225 the 2017 Nobel Prize: Thaler notes that although the field has been called behavioral economics, "It is not a different discipline: it is still economics, but it is economics done with strong injections of good psychology and other social sciences" (Thaler, 9).

CHAPTER 9

232 the West's understanding of the East: Needham's initial volume evolved into a series with coauthors, which evolved into a research institute in Cambridge that continues to publish. To date, the series spans 27 separate books, of which Needham is listed as an author or coauthor on 14.

232 the Needham Question: Years later, Needham wrote of the visit of Lu and her two colleagues, who arrived at the same time: "The fact that as scientific minds they were so much like my own raised very vividly in my consciousness the historical problem of why modern science had originated in Europe alone, and not in China or India." For a bibliography on Needham and the Needham Question, see Nathan Sivin's entry in Oxford Bibliographies; for reviews, see Finlay; Sivin.

235 who believed in his sun-centered world: Only five scholars: Westman, 309. Around the same time as Magini, Tycho Brahe, then the leading astronomer in Europe, also dismissed Copernicus's heliocentric idea publicly. He introduced his own theory of planetary motions with the title "New Sketch of a World System lately invented by the Author; In which the Old Ptolemaic Gracelessness and . . . the New Copernican Physical Absurdity of the Earth's Motion are eliminated" (Gingerich and Westman, 19).

236 complete their orbits: Mercury, 2.9 months; Venus, 7.4 months; Earth, 1 year; Mars, 1.9 years; Jupiter, 12 years; Saturn, 30 years. Uranus, Neptune, and Pluto were unknown at the time.

237 one-twentieth of 1 percent: There are 60 minutes in a degree, and 360 degrees in the arc of the sky. In "The Great Martian Catastrophe and How Kepler Fixed It," Gingerich shows how Kepler had to resolve several critical flaws of the Copernican model first, before the eight-minute discrepancy could be revealed.

237 overcame religious persecution: Personal tragedies: Kepler was abandoned by his father when young; was persecuted for his Protestant beliefs in Catholic lands; suffered through the death of his first wife and three of their children; and was forced to watch his 74-year-old mother, who enjoyed healing with herbs and potions, charged with witchcraft, jailed, and threatened with torture. (Kepler led her defense and won her eventual acquittal.)

238 by the outcome of experiments: The historian of science and leading biographer of the scientific revolution, Bernard Cohen, wrote, "The most significant development of science in the seventeenth century may have been the recognition that the laws of nature are not only written in the language of mathematics, but of higher mathematics, and that such mathematical relations must express physical causes, whose nature and mode of action are to be elucidated by the study of phenomena in relation to such causes. This 'Newtonian' aspect of modern science is now seen to have been initially Keplerian" (Cohen, "Kepler," 25). The French philosopher Voltaire put it more concisely: "Before Kepler, all men were blind. Kepler had one eye, Newton had two."

In an excellent recent history of the scientific revolution, the Nobel Prize–winning physicist Steven Weinberg wrote, "The two figures who became best known for attempts to formulate a new method for science are Francis Bacon and René Descartes. They are, in my opinion, the two individuals whose importance in the scientific revolution is most overrated." Practicing scientists prefer doers to talkers; theories that work more than statements of opinion. Kepler was a doer whose ideas worked. Bacon and Descartes were philosophers (Weinberg, 201).

238 a pace and scale of change unlike any other: For extensive discussions of the connection between the revolutions in science and industry, see H. F. Cohen; Jacob; Mokyr; Goldstone, 136–62; Lin, 22–54; Xu, and the bibliographies they provide.

239 are much more natural explanations: For recent surveys of the rise of the West debate, see Acemoglu, 45–69; both books by Daly; and Mokyr. Acemoglu summarizes arguments for the role of political and economic institutions in explaining modern disparities. For the natural experiment of Haiti and the Dominican Republic, see Jaramillo; and Diamond, 120–41, which provides many additional examples.

The debates go back to at least the eighteenth century, when the West first began to surge and many explanations were invented to explain the fact. The Goldilocks theory, for example—the notion that people in hot countries are too hot-tempered and lazy; cold countries, too stiff and sluggish—was popularized by the philosopher Montesquieu in 1748. The philosopher David Hume argued around the same time that Montesquieu's ideas were absurd, and that the true cause was the West's inherently superior race (Golinski, 175–78). Both ideas, along with variations (superior culture, religion, etc.), have persisted for over two hundred years.

240–41 at least two dozen such loonshots: A diversified portfolio of *ten* loonshots, each with a one in ten chance of success, has a 65 percent likelihood of producing at least one win because the likelihood that all ten will fail is 0.9 to the tenth power: 35 percent. A portfolio of *two dozen* has a 92 percent likelihood of producing at least one win because the likelihood that all 24 will fail is 8 percent. For the one in ten rule of thumb (in film and drug discovery), see the chapter 7 note for "*roughly one in ten*" on page 331.

242 musical chairs: Paramount was acquired by Gulf + Western, an auto parts company; Columbia Pictures was acquired by Coca-Cola, which spun it out five years later; MGM was acquired by the hotel magnate Kirk Kerkorian, then by Giancarlo Parretti with the help of the French bank Credit Lyonnais, and then went bankrupt (Parretti, described in one account as "an Italian tycoon of stunning vulgarity and shrewd charm," was convicted of fraud); Universal was acquired by a talent agency, which eventually sold it to the Japanese conglomerate Matsushita, which turned around and sold it five years later to the liquor company Seagram, which sold it five years after that to Vivendi, which handed it off four years later to GE, which merged it with NBC and then handed both to its current owner, Comcast. (Parretti: see McClintick.)

243 for rights to market it: One-off deals: For example, two small production shops optioned the rights to the book *Q&A* by Vikras Swarup. They brought in the director Danny Boyle, raised financing, and made *Slumdog Millionaire*. (Boyle based much of the film's visual story on Suketu Mehta's *Maximum City*.) Just before the film's premiere at a film festival, Fox stepped in with the cash for marketing and distribution in exchange for an equal profit share. Two days after the deal was signed—two years after nearly every studio passed—the film premiered to a standing ovation. Six months later it won the Best Picture Oscar. The structure is typical for one-off deals. The timing is rarely so good. (Roston; N. Mankad, private communication.)

243 of an industry's loonshot nursery: Not every industry has a thriving loonshot nursery, like the film or drug-discovery industries. For those that do, the larger companies, the Majors, have a choice: invest in an internal loonshot group, invest in partnering with the external loonshot nursery, or do both. There is a strong argument for doing both (they complement each other). That is the subject for a longer discussion (and a longer book).

245 It saved Leonard's life: The following year Banting was awarded the Nobel Prize for his discovery.

250 reserved for the emperor: "An imperial edict of 1004 forbade private study of astronomy and all kinds of astrological fortune-telling," with an exception for blind persons (Sun, 61).

250 they stayed dead: For example: the astronomer Taqi al-Din built a state-of-the-art observatory in the capital of the Ottoman Empire. Four years later, the sultan shut it down. A comet sighting that was supposed to have brought good news didn't (Lunde).

251 "first industrial miracle": Hobson, 50–59; Brandt, 49–50. The historian Marshall Hodgson described the European industrial revolution as "the unconscious heir of the abortive industrial revolution of Song China" (Hodgson, 197).

252 no Copernican theory: Copernicus and every scholar who used advanced mathematics in early modern Europe relied on the algebra, trigonometry, and modern numerical system developed in India and the Islamic empire and widely disseminated throughout Europe (along with the medical advances of Avicenna). The more recent challenge to the "lone genius" Eurocentric story has been the discoveries by historians, beginning in the late 1950s, of striking similarities between crucial theorems used by Copernicus and the work of the Islamic astronomers al-Dīn al-`Urdī (d. 1266), Nasīr al-Dīn al-Tūsī (d. 1274), Ibn al-Shātir (d. 1375), and Ali Qushjī (d. 1474). Noel Swerdlow and Otto Neugebauer, leading historians of early astronomy, described Copernicus as "the last Marāgha astronomer," referring to the tradition of Islamic astronomy at the Marāgha observatory (Saliba, 209). See also Al-Khalili; Lindberg, 27–167; Ragep, "Predecessors," and "Tūsī."

253 exemption from antitrust law: That was the *Federal Baseball Club v. National League* decision of 1922, which was upheld in *Flood v. Kuhn* in 1972. Both decisions are still controversial; see Alito.

254 "there would ever have been a *Principia*": "Almost certainly Newton would not have written his *Principia* had there not been a discussion by the London virtuosi of the Royal Society of the possible force responsible for the observed Keplerian motion in the planets. It was as a result of this discussion (by Hooke, Wren, and Halley) that Halley went to Cambridge to see Newton and to explore this topic with him. The subsequent encouragement of Newton by Halley and the approbation of the Royal Society were significant factors in pushing Newton to complete his researches and write them up for publication under the Royal Society's imprint. It is doubtful that without the Royal Society there would ever have been a *Principia*" (I. B. Cohen, *Puritanism*, 72).

AFTERWORD

262 idea that tumors were caused by viruses was popular: Today we know that the vast majority of cancers are caused by the accumulation of genetic mutations that are unrelated to viral infections (the toxins in cigarette smoke, for example, damage the

DNA in cells lining the lung, which can trigger lung cancer). Well-studied exceptions include infection with human papilloma viruses (HPVs), which can increase the risk of cervical cancer; and hepatitis viruses (HBV, HCV), which can increase the risk of liver cancer.

262 treating RA sells just over $30 billion annually: The TNF-alpha inhibitors including Enbrel, Remicade, Humira, and Cimzia.

263 also began as a sustaining innovation: Christensen; King; Lepore.

INDEX

Numbers in *italics* refer to figures. Numbers followed by "n" refer to endnotes.